At the age of six, when aske  she wanted to be when she gre , 'I want to travel round the wo e is the author of four collections ch, *Words the Turtle Taught Me*, ean species, emerged from her residency with the Marine Conservation Society and was shortlisted for the Ted Hughes Award. In addition to her ongoing writing residency with the British Animal Studies Network, facilitated by the University of Strathclyde, she has shared her work on BBC Two and Radio 3, and enjoyed a four-year stint as one of the poet-performers on Radio 4's *Saturday Live*.

Praise for Susan Richardson

'Spirited and varied . . . Richardson cuts to the heart of things . . . Honest and tender and makes for rich reading'
Abi Andrews, *Caught by the River*

'Cut and precise, archaic and innovative, transcendent and in-the-moment, [Susan Richardson] sees the life of the sea as a mirror of ourselves, and vice versa: always changing, always the same . . . Vital, glorious and salutary'   Philip Hoare, author of *Leviathan*

'[Richardson] writes in prehensile language, capable of grasping something vast, ancient, chthonic: the Earth in must'
Jay Griffiths, author of *Wild*

'[Richardson] shows how art and writing can furnish both beautiful and challenging reflections on our relationship with animals . . . Precise and playful; rhapsodic and rebellious'   Andy Brown

'Sparkling'   Margaret Elphinstone

'Richardson's voice swoops – passionate, ribald, funny, fierce – taking you up, on exhilarating flight, out from the cage of the everyday mind'   Eleanor O'Hanlon, author of *Eyes of the Wild*

'[Richardson] balances observation with "unseeing", science with shamanism and myth'   Chris Kinsey

'Susan Richardson beautifully marries the landscape of the polar regions with their – and her own – emotional topography'
Sara Wheeler

Also by Susan Richardson

Words the Turtle Taught Me
Skindancing
Where the Air is Rarefied
Creatures of the Intertidal Zone

Where the Seals Sing

Susan Richardson

WILLIAM
COLLINS

William Collins
An imprint of HarperCollins*Publishers*
1 London Bridge Street
London SE1 9GF

WilliamCollinsBooks.com

HarperCollins*Publishers*
Macken House
39/40 Mayor Street Upper
Dublin 1
D01 C9W8
Ireland

First published in Great Britain in 2022 by William Collins

This William Collins paperback edition published in 2023

1

Excerpts from 'Conversation' from *Treading Fast Rivers* by
Eleonore Schönmaier appear on pages 136–7 by
permission of McGill-Queen's University Press

A catalogue record for this book is available from the British Library

ISBN 978-0-00-840457-4

Typeset in ITC Garamond LT by Palimpsest Book
Production Ltd, Falkirk, Stirlingshire

Printed and bound in the UK using 100%
renewable electricity at CPI Group (UK) Ltd

MIX
Paper | Supporting
responsible forestry
FSC™ C007454

This book is produced from independently certified FSC™ paper
to ensure responsible forest management.

For more information visit: www.harpercollins.co.uk/green

For the Atlantic greys

Contents

PROLOGUE

Suckle

I hear it long before I see it. A series of cries, like the stuttering whines that precede a baby's full-blown wail. Pause. Babble of waves. Shingle-shift and drag. A snatch of silence, then another swell of cries crests and eddies.

Having hiked down hawthorn-bordered farm tracks and through pasture bristling with thistles and gorse, I've emerged onto a swerve of vertical cliffs above the bay of Aber Felin. Late summer butterflies – a ragged red admiral, a tatty speckled wood – lurch between spikes of purple loosestrife, while swallows hurtle overhead, snagging gnats.

I pick up a path at the brink of the cliffs and follow its wriggles and kinks around the semi-circle of the bay. Way below, the agitation of crags and splintered-off rocks is tempered by the presence of three pebble beaches, each differently shaped and sized, and inaccessible except by boat. I'm overlooking the first, and longest, beach now, stepping off the path into a tangle of bronzing bracken so as to better lean over the edge and spy what I hope is the origin of the cries. The geography of the bay, though, often makes pinpointing sounds difficult – they seem to bounce off the face of the cliffs and relocate in strangely amplified ways. For a moment, I think I spot a smooth, grey head breaching the water's surface but my binoculars confirm it's just a lobster pot's marker buoy.

They confirm the existence of a new tideline of debris too – a wedge of polystyrene, snarls of green fishing line, a mangled sandal edged with plastic shells.

Onwards, along the narrowest stretch of the path, with sheep confined by barbed wire to my right and the unfettered tumble of sea to my left. I make another stop, this time above the second beach, little more than a triangle of pebbles between two mini-promontories, and glimpse a skim of black and white, a streak of orange beak – an oystercatcher winging above the waves.

'Afternoon. Lovely day. How far to Strumble Head?'

A white-haired woman in shorts, her shins stippled with midge bites, is pounding towards me, powering herself along with a pair of walking poles. A man, his face and neck glossed with sweat, is plodding in her wake. While she waits for him to catch up, she reveals that they're on honeymoon and are spending it walking close to two hundred miles, the whole of the Pembrokeshire coast.

'We both love the outdoors,' the man assures me, peeling off his baseball cap and wiping forehead with forearm. 'Birds. Wildlife. All that sort of thing.'

'Have you seen any interesting wildlife?' I ask, wondering if, like me, they've been gripped by the whimpering cries.

He frowns and shrugs, dislodging his already lopsided rucksack.

'No,' the woman calls back over her shoulder, 'we've seen nothing but slugs.'

As the man tramps off in puffing pursuit again, I wander on, sampling a few blackberries en route. When another succession of cries pulses round the bay, anticipation scuds through me, but with an undertow of uneasiness too – so uncannily human, so apparently pitiful, they kindle a primitive instinct, make me want to meet the need they seem to be expressing, find a solution it's beyond me to provide.

Chivvied along by a stonechat, wing-flicking and flitting from fence post to gorse, I sidestep down a steep incline and start to curve north. Soon the final beach will come into view, the most sheltered of the three, nestled in the lee of an outcrop, free from the force of prevailing wind and waves. First, though, my gaze, as if trapped as the incidental catch of a gillnet, is dragged back towards the sea. I count two, three, four marker buoys . . . and is that a fifth? No – it resolves, through binoculars, into a mottled head that slopes to a nose fringed by a quiver of whiskers. An Atlantic grey seal, an adult cow, her large, dark eyes fixed landwards.

The cries are closer here but a beach boulder greened with algae is still concealing their source. Fortunately, I'm able to veer off the coast path, thrash through brambles and hoick myself onto a ledge, over the edge of which I can hang, stomach down on rock and patchy grass, to check out the beach from an unobstructed angle.

And there it is.

White fur stained with blood and yellow fluids. Its whole body jerking with the effort of each cry.

A recently born seal pup.

Its mother, resting a few metres away, is ignoring it, her pebble-and-kelp surrounds also smeared with blood. Another few metres further on, a mob of great black-backed gulls attacks the afterbirth, pulling it out of its steak-like shape into a stringier form, gorging torn-off scarlet strands, then tug-of-warring with the remains.

The pup appears to make an attempt to move towards its mother but manages only to flop onto its side, exposing a pink inch of umbilical cord worming from its belly fur. Mother–pup bonding seems to have been neither instant nor straightforward – the cow is more focused on the gulls than her pup and whenever a yank on the afterbirth brings one of them too close, she lunges, neck stretched, teeth bared, snarling.

Again, the pup lets out a yowl – most newborns have weaker voices and don't immediately use them as they're still feeling the benefits of the nutrients that have reached them via the placenta, but this one is plainly craving its first feed. Having managed to banish a couple of the gulls, the cow seems ready to respond at last, shunting herself towards her pup and sniffing it deliberately, thoroughly, nose-to-nose. Once familiar with its smell and henceforth assured of recognising it, she strokes its head with her clawed front flipper. It flinches and tries to move away, like a child who's reluctant to have his hair mussed. Willing her on, I watch her finally assume a feeding position, rolling onto her side, displaying more birthing blood around the base of her rear flippers, and positioning her lower belly level with the pup.

It doesn't seem to know what to do, though, and just nuzzles around vaguely in the vicinity of her body. She tries once more, shifting position until the region of both nipples is directly in front of the pup's face. I adjust my position too, try to stop the rock I'm lying on digging into my ribs. *Come on. You're almost there. Feed!*

Just as it seems to be on the cusp of latching on, the seal that I spotted in the water hurls herself out of the shallows to leave a damp track like a giant snail trail on the beach. The pup's mother, her space once more invaded, lunges afresh with her telescopic neck, elongating the smoky-black blotches that uniquely pattern her fur. Both seals let out a howl and flap their front flippers, a classic keep-away gesture, before the second cow lumbers off towards the rear of the beach.

Go back to your pup now. Please.

Rattle of pebbles as the cow resumes her feeding posture. Another few minutes of aimless nosing around by the pup. Then, having opened its mouth as if to start bleating again, it homes in on the upper nipple and begins to suck.

Reckless with relief, I stop gripping my binoculars quite so tightly, unclip the lens cap from my camera and inch a little further over the cliff. Thanks to the high fat content of Atlantic grey seals' milk – up to 60 per cent compared with humans' 4 per cent – this scrawny newborn, now that it's suckling, should double its birth weight of fourteen kilos in just over a week. Rapidly expanding into its baggy folds of skin, it'll lose the definition of its neck and take on the shape of an outsize rugby ball. By contrast, the demands of feeding will leave the cow severely blubber-depleted – she could lose up to sixty-five kilos over the lactation period – and, after seventeen to twenty-one days, having abruptly weaned her pup, she'll head out to sea to nourish herself, stomach shrunk, hips conspicuous as litter.

Photos and notes taken, I belly-crawl backwards off the ledge, scramble down to bramble level and rejoin the coast path. As I do so, I can still hear the tug and gulp of the suckling pup interspersed with the lap of waves.

A woman with a polka-dot bandana round her head and muddied trekking sandals on her feet is peering down at the beach. 'Hallo!' she says. 'Very nice, this place!'

'Have you seen the seal pup?' Thanks to the manoeuvres of both pup and cow, the algae-coated boulder is no longer hiding the former from view.

'The puppy – yes! At first, I think it is a white stone but then I see that it moves.'

She sounds Germanic and I'm aware this is my cue to embark on a polite *Where are you from? Have you been to West Wales before? How far are you walking?* conversation but can't stop myself from launching straight into a seal-spiel instead. 'It's the first pup of the season – the first to be born here at Aber Felin, anyway. Ten days earlier than the last few years – I usually see the first pup in early September – so I really didn't expect to find one today. I must have only missed the birth by twenty minutes or so. I've always wanted to watch a birth but seeing the first feed is still very special.' I'm burbling like surf and half-expect the woman to make her excuses and trek on but she seems willing enough to listen, albeit fortified by the energy bar she's extracted from her rucksack. 'It's always a bit tense, waiting to see if the cow bonds with her pup and it starts to suckle. They have to deal with all kinds of hazards, especially in the first year of their lives – as many as half may fail to survive – so it's extra-important that they get a good start. This one took a while but it seems fine now. Hopefully, it'll soon be thriving.'

The woman unscrews the lid of her aluminium bottle. 'You are very happy. Very – how you say – proud, yes?' She washes down her snack with a few swigs of water and smiles. 'You speak, I think, just as if it is *your* baby.'

Her words inspire a smile in return at the depth of my immersion in the grey seal world. And we watch the suckling pup unfurl its hind flippers like the unfolding of the opening of a story.

1

Hook-nosed Sea Pig

Rudolph's allowed to come into my bedroom but the weird, bearded man in red must keep away. I've made Mum and Dad promise that when he arrives at midnight, they'll tell him to leave any presents outside my door, or better yet, downstairs, so they can ferry them to the bottom of my bed instead.

My current clutch of worries extends way beyond preventing Father Christmas from entering my room. I'm anxious about going to school and speaking up in class. I'm anxious that I'll choke on blackcurrant Spangles. I'm anxious that Dad will die. That Mum will die. I'm anxious about falling asleep every night in case I fail to wake up in the morning. I'm anxious that I won't ever get a puppy even though I've asked for one for Christmas two hundred and fifty-six times.

I wake at just before 3 a.m. Can see my illuminated owl clock, the eyes of which move from left to right, right to left as it tocks and ticks. Relief that I've survived half the night. Relief that I don't have to fret any more about Father Christmas's arrival time. It looks like he's left a few presents, though – I can make out three un-familiar shapes on the Wombles rug at the foot of my bed. Wish it wasn't too early to start unwrapping. Wish I didn't have to go through my falling asleep routine again to try to make it through till six. Wish Dad was awake so I could ask him to read to me like he does every night at bedtime.

Could read to myself instead or write up my diary. But my diary's got to be filled in properly in cursive writing with my fountain pen, the end of which I've chewed till it's cracked right through to the cartridge of ink. And I've run out of pink blotting paper.

Maybe if I sit on the floor next to my presents it'll stop me feeling so scared about everything. Only to look, of course, and maybe for one quick touch.

Book shape.

Box shape.

The third one's different – much softer. No corners, all curves. The slope of something.

It finds its way onto my lap so I can poke it a bit and – *uh-oh* – my finger breaches the wrapping. Beneath is what feels like fur. A second finger accidentally joins the first, making the rent in the wrapping paper bigger. Not a puppy but the next best thing – a cuddly toy. What kind? My eyes can pick out two types of grey and not just because the darkness has killed all other colours – there's definitely a paler grey background and splodges of almost-black.

Can't wait – don't stop to think – tear it open. Wrapping tossed aside and the new toy cradled in my arms. I stroke him from his head all the way down the slant of his back to the flaps of his flippers. Run my finger and thumb along each spiky whisker. Hug him to me. Feel the plastic nub of his nose against my chest.

Delight quickly shifts to disquiet. I'm not meant to have opened anything till six. And not till Mum and Dad were here to see it. Try to wrap it up again? No, the paper's in pieces and trying to patch it up won't fool Mum for a minute.

Hide it then, and hope she'll forget she ever brought it to my room. Stuff the paper under the Wombles rug. And the seal? In my bed. Not at cuddle level but further down, deep in the static and slide of the nylon sheets.

Seal fur against my toes distracts me from my usual pre-sleep worries. I rock him between my feet, pretending it's the sea. Consider what to call him. Paddle in the shallows of a doze.

It's chlorine that's causing my eyes to simmer with tears, not fear. I clench my jaw so she won't know my teeth are chattering. Start to sidle towards the steps, though sidling's not easy in water that's shoulder-deep.

'Where d'you think you're going?'

This is the second Wednesday morning in a row that my class has been taken to the swimming baths. On the bus, I sit on my own, trying not to listen to the whisperings; withdraw to the corner of the changing room with its stench of Sunsilk and spite. I've just been moved up a year at school but though my brain's apparently advanced for its age, the rest of me's lagging way behind. While everyone else is mastering dives and butterfly, I'm still a shivering beginner. Today, like last week, I've clung to the edge, splashed and flailed, failed to trust the buoyancy of the float.

'You're going nowhere till you fetch this.'

The instructor chucks a black weight into the pool. It comes to rest on the tiled bottom a few metres from my feet, its underwater outline warped like one of Dalí's melting clocks. No way will I be able to pick it up by plunging my head under the surface. I'll gulp down water. Forget to hold my breath. My lungs'll overflow. I'll drown.

'You're going nowhere,' she repeats.

Tears sear my eyes again. Swallow. Breathe. Remember all the nights of the past week, when, after begging Mum to write a letter that'll get me excused, I've tried to lose myself in the pages of Dad's animal encyclopaedia. Remember all the pinnipeds, which means 'wing-footed' – the thirty-three seal species that spend part of their lives on land but are ace swimmers, appearing to fly as soon as they enter the water.

'Well?' The instructor paces the pool edge. 'I'm waiting.'

Northern elephant seal with a nose so long that it sometimes catches in its teeth.

Hooded seal with an even weirder nose, a bright pink balloon of a membrane that protrudes from one nostril to gain attention in the mating season.

Harp seal pup, moon-eyed huddle of white, doomed to be clubbed to death on the ice by Canadian hunters.

'You won't get back in time for lunch if you don't do what I've said. And if you're not in school this afternoon, how're you going to explain it to your teachers?'

While I'd willingly miss lumpy mash and luncheon meat, I don't want to miss English – we've got to write a composition today called My Most Treasured Possession.

I do my tentative, shuffling wade towards the weight, nudge it with my right toes.

Britain's got two seals – the common or harbour, and the Atlantic grey. The grey's my favourite. Not totally grey at all but lovely shades of silver and black and cream. It's the biggest land-breeding mammal we've got. And it can stay underwater for over twenty minutes.

Deep breath in.

Hold my nose.

Don't bend – just squat.

Now go.

My affection for grey seals continued to build and flourish throughout my childhood and teenage years. Although, growing up in the suburb of a stark, concrete New Town in industrial South Wales, there wasn't the slightest prospect of sighting a real live one, I committed the species' seasonal cycle to memory so that if the opportunity ever arose, I'd understand what I was witnessing without any dithering or doubt.

I devoured tragi-romantic tales from Scottish and Irish folklore too, enthralled by the character of the selkie, the seal who sheds her skin and shapeshifts into a human. If a man steals and conceals her sealskin, she's forced to stay on land, ostensibly content to be his wife but internally yearning to return to the sea. In spite of the tales' dubious sexual politics, the interactions in them still seemed easier to fathom than those of my stuttering adolescence. I hauled myself awkwardly over the pebbles and boulders of teenage friendships, moving fluently through oceans of essay writing and school exams instead.

Like the hometown I left at eighteen, the location of my chosen university offered zero opportunities for spotting seals. Over time, though, this came to matter less than expected as I began to get preoccupied by different priorities and interests. Gradually, I found my niche with a group of student writers and theatre-makers, and continued to live, write and strive with them in the years immediately after university.

Our striving finally propelled us to Edinburgh for the Fringe Festival run of my play, an earnest poetic drama about biological and literary motherhood as reflected in the lives and work of Sylvia Plath and Virginia Woolf. For the three weeks of the festival, my fellow Something Permanent Theatre Company colleagues and I rented, at a monstrously hiked-up price, a micro-flat. We had a strict rotation policy for the single bed and those not in it – cast, crew, various friends, friends of friends and supporters – slept on the floor in the kitchen-diner. Every non-performance moment was devoted to promoting the show – shouting about it on the streets of the Old Town, thrusting thousands of flyers at passers-by – and blagging our way into receptions for the free cheese and wine when our food budget ran dry. At the start of the three weeks our audience numbers were in single figures, but towards the end, thanks to a rhapsodic review in *The Scotsman*, we were miraculously at capacity.

After our closing night, I opted to travel, jubilant but frazzled, with one of the cast members, Sharon, a native New Yorker, to the Isle of Arran off Scotland's west coast to wander and ponder for a few days. To get there, we had to stagger on and off two trains, a ferry and a bus, schlepping not only our backpacks but an old school desk, one of the components of our theatre set.

For the first day in months, we had no funding to secure, no marketing to catch up on, no rewrites, no rehearsals, and decided to hitchhike from the hostel where we were staying to the beach at Kildonan in the south of Arran. We knew we had to have an Edinburgh debrief but neither of us felt inclined to tackle that yet.

I plonked myself down on the sand with my back against a rock, clocked the view of two islands – one flat and close to the coast, the other, Ailsa Craig, humped and distant – then shut my eyes and let them rest on the view inside my head instead. *I've arrived*, I grandly told myself, *at a Pivotal Moment.* Edinburgh Fringe ambition fulfilled. Play a success in all respects but the financial. A year of planning for the show at an end and the future of the theatre company unclear. My own future even more uncertain. Continue to shape the vision of Something Permanent? Take up an astonishing offer of funding, just received, to research and write in Australia? Succumb to my mother's carping and find a proper job, albeit not in her preferred setting of Marks and Spencer's Ladies' Clothing?

'You have *got* to be kidding me! No *way*!'

I'd grown used to Sharon's febrile interjections – three sheep ambling along the road had set her off earlier – so didn't feel the need to stop reflecting on my options just yet.

'This is, like, unbelievable!'

Dog chasing a ball? Boat in the bay?

'Open your frickin' eyes!' she said, giving me a shoulder-shove at the same time.

At first, I had no clue what she was gesturing at. The same sand. The same sea. Clots of spume where the two met. The same scattering of rocks on the beach.

'Can't you see? Like, right there!'

Gradually my eyes and mind tuned in. It wasn't the same scattering of rocks at all. It was a scattering of rocks topped by dappled bodies. Six, seven, eight of them and a further four heads visible beyond the rocks above the surface of the water. I tried to recall all the info I'd long ago learnt on the differences between Atlantic grey and common seals. If this was Wales or Cornwall, they'd have been greys, for sure, but both species are found around the coast of Scotland. While I remembered that the grey's scientific name, *Halichoerus grypus*, translates as 'hook-nosed sea pig', and its occasionally heard informal name is 'horsehead', photos of the male's profile, with his elongated, sloping forehead and snout, always used to remind me of a bull terrier. The common seal has more of a snub nose and cat-like face and a shorter body than the over-two-metre-long grey.

What else? Well, as far as fur patterns go, the grey seal has large inkblot splodges while the common's is finely spotted like poppy seeds sprinkled on a bap. Common seal pups are born with this coat already grown, while grey seals enter the world mantled in white fur known as lanugo – a throwback, as regards camouflage, to the height of the Ice Age – which they moult from around eleven days old in the week before they're weaned.

Though there weren't any pups, either white or speckled, on the beach, every adult's face was helpfully turned in our direction – every Roman nose, every double chin, each big splotch patterning their fur. Greys, then. Definitely greys. Thanks to Sharon's increasingly strident attempts to gain my attention, they were all hyper-aware of us. We were watching and being watched, yet their gaze seemed neither wary, as with most British wildlife encounters, nor bearishly predatory. And while it could have been attributed to Fringe Festival sleep deprivation, I had the strangest sensation of both recognising and being recognised even though this was the first time I'd seen seals outside the pages of a reference book. Somehow, their gaze felt both familiar and exotic, human and other, with the power to wisely scrutinise whatever life choices I might manage to make.

A ninth seal surfed onto the beach, then shook its head and upper body like a dog post-swim, blizzarding water. And one of the hauled-out seals, its rock washed by a wave, simultaneously lifted its rear flippers and head, mirroring the shape of my smile.

Having resolved to say yes to the opportunity to research and write in Australia, I interspersed my whirl of packing, prepping and goodbyes with spurts of research. Stirred by my epiphanic experience of the grey seal's gaze, I craved to learn more about the human–seal connection. I read that, traditionally, those who made sustained eye contact with them often claimed that seals were the souls of drowned sailors, or else they were presumed to be fallen angels, who, expelled from heaven yet not sinful enough for hell, plummeted into the sea. With their propensity for moving between ocean and land, grey seals were sometimes seen as go-betweens, inhabiting liminal spaces, even symbolically criss-crossing the border between life and death.

For the next few years, my own go-betweening was relentless. First, the move from the UK to Australia. Then, when the project for which I'd won funding came to its designated end, from poet-dramatist to writer of whatever paid the rent. From Adelaide to Brisbane. Brisbane to Hobart. Australia to New Zealand. Writer to tutor. New Zealand to Canada. Visa to visa. Loophole to loophole.

And even when I returned – grudgingly, uneasily – to the UK, I remained in motion, having pieced together a stupidly demanding, but financially essential, creative writing teaching timetable. Most days I shuttled back and forth between Swansea, Cardiff and Bristol, often arriving for my next session with just seconds to spare. Whenever I wasn't on a delayed train or in an airless seminar room, I'd be glued to my computer, tutoring online writing students. Unsurprisingly, I developed a gastrointestinal disorder and spent any non-working evenings hauled out on the sofa, clutching a hot water bottle to my belly to quell the cramps.

I finally reached peak tiredness in Dublin, where I'd managed to wangle a booking to run a weekend writing workshop. Yet even though, post-teaching, a rest in the B&B where I was billeted would have been best, I still contrived to take a bus west, then hired a bike and cycled wester.

Atlantic waves galloping shorewards. Clichéd cottages, whitewashed and thatched. Drystone walls. Fields as green as innocence.

I paused at a cove to eat a soda-bread scone and calm an agitated Spanish hiker who'd been attacked some miles back by a donkey.

Snatched sight of a bull seal and a snorting expulsion of breath as he prepared to dive again.

Though I'd pedalled into selkie territory here, I was aware that seal stories far less appealing than the mythical had emerged from the west of Ireland too. In 2004, on one of the Blasket Islands off the coast of County Kerry, over fifty seal pups were slaughtered, a brutal manifestation, it was widely believed, of the resentment harboured towards seals by much of the North Atlantic fishing community. I remembered the shock I felt when it first dawned on me that the human–grey seal relationship isn't just one of mutual curiosity and enshrining in myth. In response to the human-inspired decline in fish stocks, grey seals had been persistently scapegoated and demonised – and over the years, in Britain as well as Ireland, there had been culls, both guerrilla and government-sanctioned.

Half an hour later, at the next beach, sheltered but steeply shelving, I dug out my phone from my backpack. There'd been no network coverage all day and I was starting to feel a bit edgy about finding accommodation for the night. But here, as one, two, three seals randomly surfaced, my screen revealed one, two, three signal strength bars.

And six missed calls from my father.

In recent times, as I'd travelled ever more widely, ever more obsessively, fighting back against the phobias that paralysed me as a child, my mother's life had become ever more circumscribed. Rheumatoid arthritis had restricted her to the house, to the ground floor, and latterly to just two rooms. On the rare occasions when she agreed to see her GP, she hid – from pride? from shame? – the full extent of her struggles and pain. My father, her full-time carer, albeit that he was forbidden to use that word, likewise resisted outside help of any kind.

Today, though, in a grave tone unrecognisable from the voice that had animated so many bedtime stories, he explained she'd been taken into hospital. Her arthritis complications had progressed to the point where they couldn't be hidden any more. Like a seal's blotchy fur, her legs were now mottled with inflamed blood vessels and pressure sores.

Breathing difficulties.

Oxygen therapy needed.

Serious heart disease.

The three seal heads in the water arranged themselves into a row like the dots of an ellipsis, replacing all the words I was patently failing to find.

At last, my wind-burnt lips formed questions. What? When? How?

And an answer.

'Yes. Of course I'll come home.'

When, after several grief-clotted years following Mum's death within days of her admission to hospital, I moved to a cliff-top cottage in Pembrokeshire, desperate to embed myself in just one location at last, I had no idea there were grey seals nearby. Slowly, incrementally, by dint of hours, days, weeks of patient observation, I have become acquainted with their behaviour, grown attuned to the rhythm of their year. My own year has correspondingly fallen into a more predictable pattern. Thanks to a far less frenzied freelance work agenda, this will be the third autumn pupping season I've monitored at Aber Felin and I regularly come here at winter haul-out and moulting times too. No longer so rabidly itinerant, I've relished dropping anchor in my refuge of a bay.

Having finally finished suckling, the first-born pup of this season stretches its rear flippers again, as if flipping open a foldable fan. After the German hiker, who joked that I was speaking of the pup as if it were my baby, walked on along the coast, I couldn't resist clambering back onto the ledge from the path and bellying-down for another spell of watching.

Brimful of milk and no longer impelled to keep yowling, the pup rolls onto its side, wriggles around on the pebbles a little, then falls asleep, just as I knew it would. Yet there remains so much I still don't know about grey seals – I've done random bits of reading but there's so much else I want to learn. Connected though I now am to Aber Felin, I've started to develop a curiosity about Britain's other breeding colonies. The species is globally rare – it's said there are fewer grey seals in the world than African elephants – and confined to the Baltic, the Northeast Atlantic, including Iceland, the Faroes and the North and Barents Seas; and the Northwest Atlantic from Massachusetts to Labrador. The UK, however, at the hub of the Northeast Atlantic group, hosts almost 40 per cent of this world population. Where precisely are they all to be found? I'm aware that pupping progresses clockwise around the coast, starting in the caves and coves of Cornwall and Wales from late August, throughout Scotland's archipelagos from October and on eastern England's sandy beaches from November, but how do these breeding sites compare with Aber Felin? Who, if anyone, watches over them? Are other enthusiasts absorbed in the rhythm of grey seals' lives like me?

I switch my focus from pup to cow. She, too, is lying on her side, seeking sleep, but can't permit herself to relax completely – every so often she raises her head and glances around the beach to check that no threat to her pup is imminent. Many of the most severe hazards that pups, and indeed seals of all ages, are now facing, however, are not in the cow's power to avoid. Over the past three years, I've seen an increase in seals entangled in lost or abandoned fishing nets, known as ghost gear, for example, and there are less visible perils such as toxic pollutants.

Grey seals seem to have unaccountably surfaced at, and soothed me through, many of the transitional phases and testing stages of my life, and though it sounds fanciful, I've started to feel that I owe it to the species to become more thoroughly acquainted with the human-induced threats to which they're exposed. An idea of travelling, albeit less frenetically than before, to some of Britain's grey seal hotspots has started to form, with the aim of unveiling the many ways in which seals' welfare is impaired.

I lie on the cliff watching mother and pup till the tide turns its mind towards rising, a jag of rock jabbing my hip like a trip nagging to be taken.

2

Sanctuary

'We're not supposed to have favourites among the residents, but we do. And Ray's always been one of mine. Some of my very best memories here are with Ray.'

I'm standing at the side of a large, rectangular pool which, with its blue-tiled rim and concrete surrounds, has the look and feel of an old-fashioned lido, except that the swimmers to whom I'm being introduced are not human but seal. Next to me, dressed in a boyish uniform of turquoise T-shirt, shorts and wellies, is Dan Jarvis, one of the animal care workers here at the Cornish Seal Sanctuary. Earlier, when we first met, he swapped the broom he was carrying from his right hand to his left and shook mine with unexpected delicacy. 'So what d'you want to know?' he asked, a question so broad as to leave me momentarily flummoxed. *Everything*, I wanted to say. *Everything you know about grey seals. Just tell me everything.*

So far, he's offered an engaging overview of the personalities and medical histories of the pool's six resident grey seals who, for a variety of reasons, can't be released back into the wild. Atlanta, currently scratching her head with her front flipper while hauled out on the concrete at the far end, was rescued from a beach in the Scottish Highlands: it's believed she'd been attacked by gulls as her eyes were badly damaged, leaving them cloudily bluish and blind. Genial, elderly Lizzie, also blind and resting on her side next to Atlanta, is easily recognised by the fact that her mouth's always slightly open with her tongue tip poking out. Keeping her distance a good twenty metres away is antisocial Snoopy, a refugee from Whipsnade Zoo. The small juvenile alternately playing with a water spout and swimming in and out of a lifebuoy toy is Badger, a melanistic seal with an excess of black pigment in his fur and skin, whose underactive thyroid requires daily medication. Partially sighted Sheba, rescued back in the 1970s, is now thought, at over forty, to be the oldest captive grey seal in the world. And, finally, the seal whose head is visible just above the surface of the water near the gate where his care workers enter the pool area, is Ray.

'The shape of his head's a bit different from the other seals' because of the injuries he's suffered,' says Dan, describing how he was found in 2001 at the back of a beach with his nose pressed up against a rock face, conceivably storm-battered, at just three weeks old. 'And because of his brain damage he often does quite unusual things. Rocking horse impressions. Roly-polies round the pool. If you catch Ray on a good day, you see all sorts of things that you never see another seal doing.'

I recall my arrival at the sanctuary a few hours ago, early so as to beat the end-of-summer gush of visitors. One of the animal care workers was cleaning the pool area, alternately scrubbing the concrete and scratching, with her broom, the seal that was shadowing her.

'Yeah, that would've been Ray,' says Dan with a smile. 'He likes being dunked in the water too. If you push him down with the broom, he'll let himself float back up to the surface and he'll want you to do it over and over again. And he loves the hosepipe – he bites into it and makes the water come out everywhere if you don't pay him enough attention when you're pool-cleaning. We're always having to buy new hoses because of Ray.'

I watch Ray hoick himself out of the water, gaining leverage with his front flippers, then launch himself onto Sheba who's drowsily lounging poolside.

'His condition's progressive,' Dan continues. 'He used to have some vision but in the last few years, he's gone completely blind, which has knocked him back a bit. He became a lot more subdued, a lot less confident. But he recently seems to have got over that and is coming out with some of the crazy old behaviours that he always used to.'

I usually balk at visiting animals in captivity but am already feeling a real fondness for this unorthodox poolful of waifs and exiles. Having decided that my journey to grey seal hotspots will take me clockwise around Britain's coast, beginning in Cornwall and ending in Norfolk, mirroring the progression of each autumn's pupping season, I figured that the Seal Sanctuary would be the ideal starting point, affording me the opportunity for much closer observation than I experience at Aber Felin. Though the long-term residents might exhibit behaviour quirks unknown in their cousins in the wild, I'm able to absorb the clawed curve of front flipper, the dappling and overlapping of individual strands of fur, and the back flipper rhythm of five digits interspersed with webbing that are characteristic of all grey seals. I'm close enough to discern even their tiny ear holes, one of the principal features that distinguishes seals from sea lions. Taxonomically, the thirty-three pinniped species[1] are divided into phocids, the family of which the grey seal is a member; otariids, comprising the sea lions and fur seals; plus one odobenid, the tusked and almost hairless walrus. Only the otariids have external ear flaps and the phocids are often referred to as 'earless seals' as a result.

The sanctuary, or its early homespun incarnation, was set up in 1958 in the beach café garden of Ken Jones in St Agnes on Cornwall's north coast. Initially, he cared for the pups he rescued through trial and error and with varying degrees of success, battling to replicate the high fat content of their mother's milk with a mix of cow's milk and margarine, housing them in his caravan and a spare chalet, paying for everything from fish to vet's fees out of his own pocket. He progressed to constructing a series of pools in his garden and developed especially rewarding relationships, characterised by affection and play, with Simon, a seal with lung congestion, and Sally, who'd been blinded by the *Torrey Canyon* oil spill. Though Simon experienced several years of solicitous sanctuary living, he ultimately suffered a rapid decline in his condition. Within a week of his death, the otherwise healthy Sally also died – the vet could find no other explanation but that she'd suffered a broken heart.

Jones's book, *Seal Doctor*, details the challenges he faced in the sanctuary's first decade, from storm-lashed, cliff-side scrambles with fifteen kilos' worth of rescued pup in his arms to a catastrophic season of multiple seal deaths from septicaemia, perhaps due to the detergents that were used to clean the beaches of oil post-*Torrey Canyon*. Eventually, feeling that a larger site with at least three pools, a filtration plant to keep the water clear and a hospital was now essential, he sold his beach café, confronted local opposition, endured a two-year struggle to get the plans agreed and finally succeeded in opening the Cornish Seal Sanctuary in its current location.

As I've discovered, this new setting is in the far southwest, in the heel of the boot that's the shape of Cornwall, on the Helford River. Though it's approached through a residential area on the outskirts of the village of Gweek, a name that sounds remarkably like the cry of a newborn pup, the sanctuary itself feels pleasantly rural, with views over the river and low hills like the rounded mounds of a seal haul-out.

By the late seventies, having spent twenty years of his life tending to seals' needs for up to eighteen hours a day, seven days a week, Jones claims a deep connection – 'I know their thoughts, their feelings, their sufferings.' I suspect he's only half-jesting when he confesses that on the occasion of his silver wedding anniversary, he and his wife, Mary, 'had a little party and invited all the seals'.

More than two thousand have been rescued in the past six decades, although the sanctuary has shifted in character, as well as in physical distance, from its unsophisticated beginnings in Jones's back garden caravan. In 1993, it was acquired by Merlin

Entertainments, one of the largest leisure enterprises in the world, with attractions that include Alton Towers, Legoland and Madame Tussauds. As is apparent from the signs pointing to the Humboldt penguins' and Asian short-clawed otters' enclosures, the sanctuary has had to learn to reconcile the need to provide critical care to rescued seals with the obligation to concurrently deliver a 'family visitor experience'.

The scheduling of regular feeding times that satisfy both the seals' appetite for herring and the public's for entertainment is further evidence of this. While Dan and I have been chatting about Ray and co., an assortment of mothers, fathers, kids, buggies, young couples, older folk and dogs on leads have arrived, some taking the woodland walk from the visitor centre, others boarding the Safari Bus. A child raising and letting fall the pendulous ears of an excitedly panting cocker spaniel is tugged away by her father – 'You've come here to see seals, not dogs.'

Dan leads me to a second resident grey seals' pool containing 'the four boys', whose feeding is just getting underway. Though a row of visitors with sunhats on heads and toddlers on shoulders is partially obstructing my view, I can see enough to conclude that this pool, backed by natural rock, and not so rigidly rectangular, is much less lido-like. There's not a tile in sight, though its insides are painted deep blue, giving the water an incongruously tropical hue.

I watch as one of Dan's fellow animal care workers persuades the largest of the seals to heave his vast self out of the water and onto the poolside. 'Good boy!' she enthuses, rewarding him with a fish from her bucket, then encouraging him to touch his nose to a stick while she moves around him to do a quick medical check, running her hands down the bulk of him, lifting his flippers and looking at his teeth. She examines his neck, too, around which are bunched thick, corrugated folds of skin. If he were living in the wild, these would offer some protection to his throat during fights with other bulls in the breeding season. Grey seals are sexually dimorphic – the males and females exhibit different physical characteristics – as is evident from the slope of this bull's nose, broader and more pronounced than the snouts of the cows in the first pool, and his longer, heavier body. While he might be expected to top three hundred kilos, adult females tend to reach just half to two-thirds of that weight.

'That's Yule Logs. He's the dominant male here,' says Dan. 'He was very insecure in his first years at the sanctuary but now this boys' pool is very settled.'

'Oh my god – he's massive!' The woman directly in front of me backs away slightly, stepping on my toes, even though a transparent plastic screen separates us from the seals. 'I would literally poo my pants if I was in the water with that.'

Dan explains that Yule Logs was originally rescued as a pup by a marine park in 1989 but, following his recovery, they misguidedly decided to keep him on show rather than release him back into the wild. Four years later, this decision was reversed but since he'd been in captivity from such a young age, he had no idea how to feed himself. Three months on, he had to be re-rescued, severely underweight, having spent much of his post-release time chasing children with buckets and spades up and down the local beach because he associated humans carrying buckets with the provision of fish. A safe and permanent home was subsequently found for him here at Gweek.

Inside, I'm raging at the chain of unethical decisions that deprived Yule Logs of a life in the wild, while simultaneously craning to see the next seal whose story Dan is outlining.

'Flipper's the old man of the sanctuary,' he tells me, pointing towards another sizeable seal, dark-furred like most bulls but with pink discolouration of the nose. 'It used to be Flipper who was in charge but when he went past breeding age, Yule Logs took over.' Flipper provides a connection with the Ken Jones era, having been rescued back in the 1980s at two weeks old, suffering from a severe respiratory infection, believed to have resulted from his having inhaled a chemical pollutant. As well as causing the nose-pinking, the chemical had burnt his airways and lungs, leaving him only able to dive and hold his breath for five minutes, and consequently unable to successfully forage and feed. A few years back, he underwent cataract surgery, reducing the team of animal care workers to joyful tears when they witnessed Flipper's awareness that he could finally see again.

'He's incredibly patient with everyone – especially with Pumpkin . . . that's him over there,' Dan continues, gesturing towards the smallest of the seals, who's receiving his fish from an animal care worker with a whistle in her mouth on the far side of the pool. 'Pumpkin likes to be next to Flipper and touch him all the time. Flipper could easily turn round and bite him but he never does.'

I rarely have the opportunity to observe male seals at Aber Felin as it's mostly populated by adult cows and, perhaps because of my affection for my father, I instantly feel drawn to this gentle, elderly bull. Having grown a bit frustrated by the row of visitors semi-blocking my view, I worm my way to the front for a closer look, ending up next to a woman who's pressing her twin toddlers up against the plastic screen. 'Look, Liam! Look, Lucy! He knows you're waving at him!'

In my new position, I supplement what I'm learning from Dan with some reading of the signs that are attached to the pool's perimeter. They're unashamedly anthropomorphic – 'Hello, my name is Pumpkin. I was rescued in Guernsey. I was very poorly and weak when I arrived and spent a long time in the hospital.' Due to a persistent urinary infection, later attributed to kidney stones, young Pumpkin has remained at the sanctuary as a long-term resident.

With feeding time drawing to a close, Pumpkin launches himself back into the water and makes an immediate beeline for Flipper. Most of the crowd starts to disperse, drawn towards one of the food outlets by the advent of their own feeding time. I focus for a while on the pool's fourth seal ('Greetings! My name is Marlin. I like eating as much mackerel and herring as possible!') who, according to Dan, in spite of being blind, 'learns quicker than the two old boys', responding to commands through sound, physical contact and disturbance of the water. Flipper seems intent on seeking sleep now, lying full-stretch in the pool, head below the surface, in a posture known as logging, and refusing to be disrupted by the clambering of Pumpkin on his back.

Of the visitors, only a teenage couple remains at the side of the boys' pool, his arm round her minuscule waist, her head resting on his bony shoulder. 'This is, like, amazing, Josh,' she says. 'I've never seen anything wild before.'

A helicopter from the nearby RAF base wup-wups over the sanctuary as I make my way uphill to the heart of its rescue, rehab and release programme – the Seal Hospital. Because the autumn pupping season's only just got underway, however, the squat, single-storey building is almost entirely empty of both staff and seals: there's only one pup in residence. Spotted howling and hauling herself across a West Cornwall beach, trying to suckle every rock and clump of seaweed in her desperate path, behaviour typical of a pup that's become separated from its mother, she was brought into the sanctuary's care by an abseiling Dan and a colleague from British Divers Marine Life Rescue ten days previously. Her initial hospital assessment confirmed that she was just a couple of days old.

Although it's not possible to see her at present – the floor-to-ceiling window through which recuperating pups can be viewed has justifiably been blacked out to give her some peace and privacy – I can hear a low moaning that sounds quite different from the hungry pup mewling that I'm used to from the beaches at Aber Felin. It strikes me as being a more distressed, existential howl somehow. Dan, though, assures me that the pup's doing well. 'She hasn't put a lot of weight on yet but she's still making very good progress.' I'm projecting my own conflicted feelings, then – relief that she was rescued and spared from starving to death mixed with discomfort at the force-feeding by tube and isolation in an alien indoor pen that she's currently enduring.

'Every year, we take in around seventy rescued pups,' Dan tells me. 'By January, we'll be full here in the hospital and the outside pools will be packed with pups too. In the final stage of rehab, we put them in with Snoopy and Sheba and Ray so they can compete for food and build up their weight before they're released back into the wild.' Being inside the hospital has dulled the blue of his eyes but his seal-inspired smile is still as broad as the outdoors. 'You'll just have to come back in January and see!'

I'm able to stay in Cornwall for only one more day on this September visit and I'm itching to spend it watching grey seals in their natural environment. A mottled colony of disquieting thoughts has hauled out in my mind since my time at the sanctuary – thoughts about the ill-conceived decisions that led to Yule Logs spending his life in captivity, thoughts about the challenge of balancing crucial rescue work with the provision of money-generating entertainment[2] – but I'm hoping that hours on a cliff top with binoculars pressed to eyes will help restore equilibrium.

I've walked around and beyond St Ives Bay once before, years ago when I was doing research for my Virginia Woolf play. Woolf spent many childhood holidays in St Ives, and Cornwall remained a source of solace, and inspiration for both her fiction and auto-biographical writing, for much of her adult life. The octagonal white tower on the islet opposite the cliff on which I'm now walking is known to be the titular lighthouse of Woolf's ground-breaking 1927 modernist novel. And a quote from *To the Lighthouse*, from the internal monologue of the artist heroine, Lily Briscoe ('. . . making of the moment something permanent . . . In the midst of chaos there was shape . . .'), was the origin of the name of the theatre company for which I wrote. We grandiosely hoped that our Something Permanent theatre-making would create shape and stability from personal and political chaos, and posed for moody, monochrome St Ives Bay photos to use in the publicity for our Edinburgh show. Today, though, after just a nostalgic glance in the lighthouse's direction, I choose not to linger, leaving the field full of cars and campervans, walkers and dogs, to hike along the coast to the cove that Dan recommended.

Guaranteed seals, he told me.

Though in Pembrokeshire, especially now that pupping's underway, I'm used to hearing seals before I see them, it's the high-density cluster of seal-related signage that I'm most immediately aware of here. National Trust signs telling me I'm about to enter a Sensitive Wildlife Area, urging me to keep my voice down and my dog on a lead. An information board about seal-watching etiquette with photographic evidence of chilled seal behaviour versus panicky stampeding into the sea. An instruction to phone, if necessary, Cornwall Wildlife Trust's Disturbance Hotline. An enticing mention of Cornwall Seal Group and the monitoring work it undertakes. 'Shhhhh, Whisper for Wildlife' notices pinned to low wooden barriers, positioned to prevent people toppling off the cliff.

'THAT'S NOT A SEAL – THAT'S A ROCK!' yells a child in flashing trainers who has yet to take the do-not-disturb advice on board.

I inch closer to the edge and peek over. It's a dizzying distance from cliff top to sea and the composition of the beach – an area of sand plus a low-tide exposure of boulders – is a little different from Aber Felin. Some of the larger rocks are garnished with a sleeping seal, hind flippers fronding the edges like kelp. With the naked eye, thanks to the pebble-like patterns of their fur, it's not easy to differentiate between seal and stone, but as soon as I use my binoculars, the camouflage is less befuddling. Twenty-four adults, more bulls than cows, and three noticeably smaller juveniles.

For the most part, they seem relaxed, stretched out on their sides in sleep, occasionally waking to have an elaborate scratch. They're lying in random directions rather than all facing the sea, which would indicate unease and readiness for a quick getaway, although one cow seems much more anxious than the rest, persistently raising her head to look up towards the cliff top. Grey seals' vision isn't quite as efficient on land as underwater so I'm guessing she's responding more to the sound of all the shouts and barks and shrieks than the sight of the constant toing and froing. Sometimes, one of the younger seals, responding to the older cow's vigilance, glances up too, yet neither, at this point, seems impelled to bolt into the sea.

In fact, my own disturbance detector seems to have been rather more activated than that of the seals and it's not long before I decide to walk a couple of hundred metres further along the cliffs to where there are fewer visitors. There's room to lie on the grass here and look beyond the shoreline to the outcrops of rock in the mouth of the cove. Another mix of six cows and bulls is hauled out there while two young seals are swimming together. The plunging and rolling, weaving and Celtic knot-twisting of juvenile seals at play can last for hours and so can my devotion to observing them. As I watch, I ponder both on what I've experienced so far today and the broader issue of human–seal interaction. Compared with the peace and seal-viewing freedom that I enjoy when over-looking my more remote bay at Aber Felin, today feels like a domestication of the wildlife-watching process. Yet since the cove here is so close to popular holiday destinations and the seals are so at risk of disturbance, all the signs and warnings are clearly needed. Is it a seasonal concern and shall I make a return visit in winter to find out? And how else is human–seal conflict going to manifest itself in the course of my journey around Britain's coast?

My train of thought is derailed by the arrival of a man in trop-ical shorts who sprawls at the cliff edge beside me. Rather than using a lead, he has one insubstantial finger hooked under the collar of his wriggling Jack Russell and I'm now on high alert to pre-empt a squirm right over the precipice.

More shouts. More squeals. Two kids brandishing *Star Wars* lightsabers, charging along the cliff top with the exuberance of young seals. I stand, brush grass seeds from my T-shirt and begin to walk on. A final sign informs me that I'm now Leaving the Sensitive Wildlife Area.

'Thank You For Talking Quietly', it says.

3

Seal HQ

'It's really shit being a female seal. You're pregnant 24/7 and then you die.'

My curiosity having been piqued by the reference to Cornwall Seal Group on the cliff-top information board near St Ives four months ago, I've tracked down the trust's founder and director, seal enthusiast extraordinaire, Sue Sayer. I've picked one of the worst weeks of the winter for my return visit, though, with today's wind and rain the warm-up act for the storm that's due to headline tomorrow.

Joining Sue for one of her seal surveys of the cove, I've stepped over the wooden barrier intended to prevent folk from getting too close to the edge and hunkered down at the cliff's very brink. Our blustery trudge here had a very different feel from my stroll of last September. The field car park shut, with the herd of camper-vans replaced by a flock of oystercatchers probing the soggy ground. Peering at the lighthouse through a bead curtain of rain. And no one else walking the cliffs at all.

Sue's already taken several photos of the bay and is now focused on filling out the survey sheet tucked inside her waterproof clip-board cover. 'My colleague and I do a three-hour survey two mornings a week,' she explains. 'And then we go home and do five or six hours ID-ing, hoping to match up some of the seals we've photographed with those who are already in our database.'

My casual trips to Aber Felin – once a week if I can manage it, thoughts scrawled in a battered notepad – feel very amateurish in comparison, and my present attempt at speed-counting seals even more so. In just a few seconds, Sue's able to approximate the number of seals on the beach, dividing them, by eye, into groups of five. While I'm still laboriously counting and double-counting, losing track of where I've got to and having to start over, she's filled in most of the columns on her form. Forty-eight, I decide, until a recount to make sure gives me only forty-six. No – forty-eight – I was right the first time. And here's Sue with the weather and tidal conditions, number of adult males, adult females, juveniles, hauled-out seals and seals in the sea all recorded.

'How long have you been monitoring this colony?' I ask.

'Well, first of all,' she says, 'we have to question the whole concept of a stable colony. Different sites have different roles for seals at different times of the year. It's better to think of this beach as a motorway service station, with lots of seals moving through. We recorded 902 different seals here last year. Only two were identified twenty or more times, while 89 per cent – that's 803 of them – were identified fewer than five, so none are thought to be resident all year round.'

Her ability to so readily recall and relay all this information is nothing short of awesome. And I'm enthralled by her theory about the flux and flexibility of grey seal sites too. I'm aware that seal numbers at Aber Felin vary throughout the year of course but have lazily assumed I've been watching the same core group, with the same cows coming to pup each autumn and the same seals arriving to bump up the numbers during the annual moult.

'Some sites are important for juveniles,' Sue continues, 'some for males, some for females, some for pupping. This isn't really a pupping beach. We only had a handful of pups here last year – the pregnant cows prefer to go somewhere more isolated.'

Gradually, albeit through rain-spattered glasses, I begin to tune in to the beach on a more subtle level, observing not only the seals who are currently hauled out, but also picking up on some backstories. On the left, like the rakings of an erratic sand artist, wide tracks, caused by foreflipper-drag and body-bounce, indicating that several seals have recently lolloped away. On the far side, rubble and slabs at the cliffs' base, suggesting there's been a rock fall here at some stage. Such events can crush resting seals to death and, with extreme storms triggered by climate change on the increase, they're likely to become an ever more frequent hazard.

'Raised Eyebrow!' Sue suddenly exclaims.

'Sorry?'

'The seal who's just come onto the beach. I recognise her from her fur pattern. She's been seen in the Scillies too.'

'Why's she known as Raised Eyebrow?' I ask, imagining an unusual alignment of the whiskery tufts above her eyes or else some aberrant behaviour that caused consternation in whoever first observed it.

'We name all the seals after their fur patterns. For example, Snowdrop has three petal shapes on her neck. Duchess has black beads like pearls.'

I watch Raised Eyebrow have a vigorous back scratch, rolling around on the sand, but even though this affords me an assortment of views of her neck and torso, I can't distinguish the eyebrow shape after which she's been named from the myriad other black blotches that pattern her fur.

From deep within the hood of her waterproof, Sue describes, in greater detail, the pioneering photo ID work she's been undertaking for some two decades and the ever-expanding database of seals she's established, a project which remains one of Cornwall Seal Group Research Trust's key achievements. In spite of the annual moult, each seal's unique markings remain the same throughout his or her lifetime, and thanks to Sue's twice-weekly monitoring, contributions from her colleagues elsewhere in the South West, information gleaned on boat surveys and photos submitted by members of the public, hundreds of seals have been re-sighted at many locations around the Cornish coast. There have been matches over the years from further afield too, such as Pliers, who's been spotted on Skomer Island off Pembrokeshire, and, most remarkable of all, Teddy, who, says Sue, 'I recognised on a trip to France!'

Listening to her speak is making me itch to get to know the seals on the beach at Aber Felin as individuals rather than as an amorphous group. Though there have been a few I've easily identified because of distinctive scars, I've never methodically taken photos of each seal's left and right profile or tried to memorise individual fur patterns but this meeting with Sue is nudging me to start. It would be fascinating to know if any of the seals I've seen at Aber Felin move between West Wales and Cornwall too.

'Sure you don't want to go somewhere warm and dry?' she asks, not for the first time, as I struggle to get my pen to write in my semi-saturated notebook. I shake my head, causing yet more rain to cascade from the rim of my hood onto its pages.

We do, though, move to the grassy spot further round the cove where I was joined by the man with the fidgety Jack Russell in September. From here, we watch a young bull clambering on top of a juvenile cow at right angles, then rolling-pinning her body back and forth on the sand. She battles to escape from beneath him and he eventually lets her writhe free, after which they embark on some face-to-face, neck-against-neck wrestling. 'It's common to see male and female juveniles playing in this way,' says Sue. 'And you'll often see males playing with other males too. Rarely a cow playing with another cow, though.'

'Why's that?'

'Because the cows are always pregnant,' she says, returning to the topic she had a mild rant about earlier. Even as females are weaning their current pup, they're in season again, being stalked by brawling bulls and their reproductive life might last from the age of five to over thirty. 'Cows do very little in terms of interaction other than shoving off unwanted males and other cows – they're just beholden to the gene machine.'

Unlike Aber Felin, this beach is frequented by rather more bulls than cows, which has no doubt contributed to Sue's appreciation of the former. 'We can get some beautiful girlies here but there's nothing more beautiful than an adult male. You know when you put your hand in a sock to make a puppet? That's what their faces remind me of!'

I'm keen to learn more about bull behaviour since I mostly tend to see them in the breeding season when they're in mate-with-multiple-cows mode. The emphasis on this aspect of their lives in books and scientific articles often seems to imply that the only noteworthy males in seal society are these dominant bulls.

'Well, that's got to be bollocks,' says Sue. 'Someone recently sent me a photo of four juveniles and an adult male faffing about together and it looked just like a fishing lesson. My theory is that adult non-breeding males, far from not having a role in seal society, are teaching the little ones how to be a seal. Which is probably the most important job you can have.'

Right now, keeping the rain out feels like a more important job still. The wind's been driving it into any crevices that I haven't zipped or velcroed sufficiently and it's since upgraded to hail, prickly and percussive across the expanse of my back. The next time Sue suggests we move somewhere warm and dry, I happily comply.

As soon as we enter her house, Sue deposits her clipboard, binoculars, camera and mini-tripod into a chest just inside the front door, ready to be packed into her rucksack the next time she does a survey, a stark contrast with my own disorganised dash through every room to gather notepads, pens, binos and weatherproof items of clothing before I go to watch seals at Aber Felin. A storage net full of gloves and hats, again easily grabbable prior to a cliff-top watch, is also positioned just inside the door.

In the kitchen, Sue swaps her fleece for a striped apron and throws herself into the task of making pancakes. Now there's no longer the need to keep noise to a minimum to avoid disturbing the seals, she's even chattier. 'It's great to meet a fellow seal obsessive – it doesn't get any better than that!' she says, briskly whisking eggs, milk and flour.

I have a choice between a seal and a Chelsea FC coaster on which to place my mug of tea and naturally choose the former, which triggers an entertaining narrative from Sue about how her preoccupation with seals evolved. Having reached the age of forty feeling that her ambitions needed to be more clearly defined, she did some mind mapping to determine what mattered most to her. Four things emerged as central – cats, her partner, Chelsea FC and seals. It was at this point that she realised she wanted to work towards becoming a 'nationally renowned seal researcher'.

Cornwall Seal Group Research Trust rapidly progressed from its modest beginnings. At first Sue convened the group informally and infrequently here at her house ('It was just other people I'd met on the cliff while I was watching'), then when she gave up her job in education, the meetings became monthly. Several times in the course of her account she says, 'I'm going to shut up now,' miming pulling a zip across her lips, but within seconds, she's off again, eager to continue outlining the group's accomplishments. From becoming a registered charity to winning funding to write a seal entanglement report and present findings at a Marine Mammalogy conference in San Francisco, it's clear that the group is now a vital force in marine conservation, not just in the South West but far beyond.[1]

Replete with pancakes, we carry our mugs upstairs to her study, otherwise known as Seal HQ. I'm offered the office chair over which hangs a tote bag with 'Crazy Seal Lady' printed on the side, while Sue perches on a giant yellow exercise ball, switches on the PC and calls up her mighty photo ID catalogue. Later, she'll import this morning's images – each survey yields several hundred new photos – and painstakingly scroll through the catalogue for matches, but for now, she calls up some favourite seals onto the ultra-widescreen, zooming in on the fur patterns after which she's named them.

'This is Hook. He was the first seal I identified between two sites so he's very special. . . That's Trunk who spends all summer in Newquay Harbour. . . Here's Sunrise – d'you see the shapes on his neck like sunrays? . . . Now, where's Medallion Man? I've got a picture of him somewhere – come on, Sue, switch your brain on. Too excited – that's what the problem is!'

I contemplate the diligence and memory power it must take to find a match in the database, to identify the same seal twice over, and recall playing the card game Concentration as a child. Instead of aiming to turn over pairs of matching picture cards from the fifty or so that are laid face-down on the table, though, Sue has thousands of seal images and subtle variations in their markings to consider. Yet she shrugs off my wonderstruck interjections – 'Our brain spends its whole life looking for patterns.'

Her mood becomes more sombre as she introduces me to Sandy, a young female seal whom she helped to rescue. After printing me a copy of Sandy's Story, measured and factual, from a folder of seal chronicles on her PC, she narrates the more impassioned version. 'It was blowing a force six the day we rescued her – bloody freezing. We were lined up being her windbreak to stop her being buried in the sand. Somehow managed to get her off the beach and to the sanctuary at Gweek but she was completely unresponsive. Survived less than forty-eight hours. I went to the post-mortem – turns out she spent most of her life entangled in fishing gear, then was probably bycaught too. A double whammy.'

I've been expecting to confront the issue of seals' entanglements in ghost gear, as well as their incidental capture, alongside target fish species, in gillnets and commercial trawls. I haven't, though, been prepared for it to be so graphically embodied in a single, luckless seal.

While we're on the subject of anthropogenic impacts, I mention my visit of September and the level of noise disturbance to which the seals were being subjected. On an inclement, out-of-season day like today, it seemed like a far less consequential problem but

Sue leaves me in no doubt that it's ongoing and serious, instantly fishing another set of figures from her memory's depths. Disturbance was observed on almost half of her surveys last year and 111 incidents of seals bolting from the beach into the sea were recorded. Seals need to haul out for a range of vital reasons including body temperature regulation and oxygen replenishment following periods of foraging and diving. If they're not permitted to enjoy sustained periods of rest, they're unable to derive the energy they need from digestion, and, with heart and breathing rates and stress hormones all raised unnecessarily, repeated fleeing into the sea squanders essential stores of energy too. Long-term, their reproductive success and even life expectancy can be compromised as a result of this persistent physiological disruption.

Having to complete the seal-sexing quiz that Sue's designed for students at a local college causes a degree of physiological disruption for me, at least as far as adrenaline production goes. Thankfully, though, I ace it, with only the photo of the final juvenile causing me some temporary bother. The contrast in body size and nose shape is much less pronounced in young males and females than in adults and though fur colour can be a useful guide – cows tend to be paler with dark splodges while bulls are generally darker with some lighter patches – this method of determining a seal's sex isn't 100 per cent reliable.

'Yes, that's a girlie. Tricky one, that – well done!'

After spending another couple of hours in Seal HQ, I feel ever more appreciative of Sue's blend of scientific rigour and unapologetic affection for the seals she IDs, her fierce defence of their welfare and occasional winsome slips-of-the-tongue in referring to them as 'people'. There is, though, one final question to which I've yet to get an answer.

'What exactly is it about grey seals that you find so fascinating?' I ask, as I pack away my notebooks and all the leaflets, seal stories and info on other Cornish seal sites that she's given me.

'Whiskers!' she says at once. 'Their whiskers are awesome. And by whiskers I also include eyebrows – their eyebrows are very special!' When we reach the foot of the stairs, she adds, 'Their eyes too – there's a communication, a real connection.'

By the time I've retrieved my waterproof from the downstairs shower room where it's been drying under the wingspan of a hanging wooden gannet, she's thought of another reason. 'I also like that there's no such thing as an average seal. They do their own thing and that's what makes studying them so bloody hard. There are patterns and stereotypes but there are always exceptions.'

At the front door, she gives me a goodbye hug. 'Never think you know about seals. As soon as you think you've got them sussed, they'll prove you wrong. Every time.'

The next day's weather proves to be similarly tricksy. Though I'm dying to visit a few of Sue's other recommended Cornish seal sites, the storm has arrived on the north coast, bringing 'monumental tides', and waves described as 'very high, if not phenomenal'. Propelled by gale-force winds, clouds are careering across the sky like a spooked group of seals fleeing from beach into ocean.

Nevertheless, having studied my map of Cornwall and researched the projected path of the storm, I decide to risk a trip to a cove on the Roseland Peninsula: with its south coast location, it's likely to be far more sheltered than any of Sue's other seal spots. The fact that the King Harry Ferry is still being permitted to carry cars across the River Fal is an auspicious sign and as it makes its sedate way to the opposite bank, there's little hint of wind buffeting or rain.

I don't have the most explicit information on how to locate the cove where seals are supposed to congregate. Find the National Trust car park, walk to the coast, turn right and keep going till you see them. These directions are accompanied by hazy memories of navigating the Roseland Peninsula over a decade ago when, feeling limp and fragile following an episode of jaundice, I recuperated here with a week of reading, gentle ambles and overpriced cream teas.

Seal HQ

Another propitious sign. With the car parked and a track to the coast squelched along, I immediately spot the snout of a sleeping cow seal thrusting skywards from the water, the rest of her body vertical beneath the surface, a posture known as bottling. Every so often, she sinks out of sight, then, after a minute or so, her bewhiskered snout re-emerges so she can breathe a few times. I remember reading a study of the sleep of captive grey seals, in which a pattern of breaths and movements exactly akin to this cow's current sleep behaviour was described. Telemetry devices were implanted in the study animals' necks and backs that yielded data on their brain and heart activity and it was the first study to confirm that grey seals sleep underwater.

More recent research has shown that some sea lions and fur seals exhibit unihemispheric sleep while underwater, whereby one half of the brain remains awake, yielding various benefits including continuous vigilance. Grey seals don't experience unihemispheric sleep of this kind but they've nonetheless evolved their own subaquatic sleep adaptations – they never enter REM sleep when underwater, for example, but naturally do so if their nostrils are above the surface and when on land.

Leaving the cow seal to slumber on, with small waves occasionally breaking over her face, veiling it in spray, I turn right along the trail. A row of Alexanders, their yellow-green umbels emerging early in the South West's mild climate, sway in what's no more than a boisterous breeze. The path is slippery in places but otherwise it's easy walking – much more protected from the elements than yesterday, with no breath-seizing precipices over which to lean.

Finding, and engaging with, the seal cove turns out to be easier than expected too. As it comprises a sweep of sand interrupted by only a few scatterings of rock and shingle, the seals that have hauled out aren't camouflaged like on the boulder-strewn beaches to which I'm accustomed. Seventeen seals, I effortlessly estimate, twelve of whom are bulls. Three silver-grey cows, two dozing on their bellies, the other deep-sleeping in the most defenceless position of all, on her back. Two juveniles recently emerged from the sea, fur still wet, their markings as a consequence more precisely defined than those of their dried-out companions.

I pick my way down the slope from the coast path to an expanse of grass just above the beach from which I can watch for a while. It's enthralling to see so many bulls lying calmly side by side, tolerant of each other's presence: free, at this time of year, from the biological imperative of defending a collection of cows, they're completely still but for an occasional abyssal yawn or head scratch. Whereas the females with whom I'm familiar on the beach at Aber Felin are irritated when their personal space gets invaded, moaning and flapping their flippers, none of these bulls seems bothered by the proximity of their fellow seals at all.

While I watch, I extract a mini-radio from the pocket of my waterproof and plug in the earbuds. I've brought it along so I can keep up to date with the storm reports and find out if I need to make a quick getaway. Radio Cornwall's programme schedule seems to have been replaced by one long, storm-filled news bulletin, interspersed with listeners' eyewitness phone calls. It's hard to reconcile the tranquil scene in front of me with what's taking place barely twenty-five miles away on the opposite coast. Several thousand houses are without power. Twenty-metre waves have been recorded. A gust of wind has lifted up a pony's stable, transported it over a hedge, and deposited it on an adjacent lane.

It feels nothing short of miraculous that I've been able to connect with mellow bull energy today, as well as a mellow sea. The only turbulence to which I'm privy is being generated by the two juveniles. They've started mock-tussling, mounting and mouthing each other, scuffing up squalls of sand.

Once the storm has exhausted itself, leaving behind a chill, still brightness, I take the opportunity to visit one more seal site on Sue's list, probably the most exposed. Having passed 'the most southerly estate agent on the mainland', I reach 'the most southerly land lighthouse in the UK' where the road runs out. A handful of buildings lines the cliff-top path between the car park and Lizard Point including a gift shop (closed), an artist's studio (locked) and a National Trust hut (open, and helpfully advertising itself as a 'Wildlife Watchpoint', without which I would have surely walked on and not thought to scan the sea for seals at all).

Inside the hut is a huddle of Sue's seal volunteers who've taken responsibility for monitoring this particular site, developing their own ID database and regularly contributing their photos to Sue's all-Cornwall catalogue. Though it's crowded with bodies, binoculars, mammoth camera lenses and an ultra-patient dog, I also manage to squeeze inside. Terry, of whom I can see only glasses, moustache and nose, cocooned as he is in high-collared waterproof, woolly hat and hood, has been coming here almost every day for over three years, as have married couple Enid and Alec. In that period, they've managed to ID more than seven hundred individual seals, photographing them both bottling in the water and hauled out on the rocks beyond Lizard Point.

The fourth occupant of the hut is freelance ecologist Tony who controversially announces that he's not a big fan of seals. 'They smell of fish and fart a lot. And there's enough seal obsessives around the coast of Cornwall. We need to get people interested in other things like choughs instead.'

I'm not sure how seriously to take his pronouncement, for between the male members of the group there's mock gibing galore. I manoeuvre myself to the window against their backdrop of banter and scour the sea from cliff base to horizon, first with the naked eye and then with binoculars. In spite of the lack of wind, each post-storm wave is frothing and roiling and the rocks, currently spotlit by sun, on which the seals customarily haul out, are empty.

To compensate, Terry fishes out a bundle of faded photos of some of his favourite seals. 'This is Archer Lady, our matriarch. This is Mister and Missus. This is Black-Eyed Susan. . .'

'I named that one,' Enid chips in.

'And this is Key, who had her pup just round the corner – two years in a row!'

While I offer admiring comments on the rest of the photos, Alec returns from his short walk to the 'most southerly café in the UK' with cups of coffee for us all. Between sips, I hear how they became involved in seal surveying via spending time here monitoring choughs. This acrobatic, red-billed corvid became extinct in Cornwall in the second half of the twentieth century but recently started to make a comeback – breeding pairs are now in double figures and one of the nesting sites is here at Lizard Point.

'Can we have some of your choughs, please?' interrupts Tony in a manner that suggests I'm being selfish by keeping so many all to myself in West Wales. 'We could do with some of your Manx shearwaters too.'

Having finished her coffee and perhaps seeking a break from the constant joshing, Enid pulls on a stripy hat, zips up her waterproof and suggests that she and I leave the hut and look for seals a little further afield. After a quick glance down at Polpeor Cove to our right, overlooked by a former lifeboat station building, the tiny beach blanketed with kelp, we start wandering east along the coast path past the lighthouse.

As the tide falls, ever more rocks are emerging, each one containing the possibility of bearing, or being, a seal. 'I know every one of these rocks,' says Enid, intimating that she's moved way beyond the point of confusing live animals and gneiss, but in spite of her conviction, I still sweep my binoculars from one crag to another in case I'm missing something.

'Neil! Neil!' suddenly yells a woman just ahead of us on the path. 'There's a seal!' She does some exaggerated pointing and waving while her male companion squints hopelessly into the sun and their poodle gambols around, barking.

'It's not a seal, it's seaweed,' mutters Enid. 'But she's so excited I'm not going to spoil it.'

Suspecting that any seal in the vicinity would have dived out of sight and sought to put as much distance between itself and the shouty excitement as possible, we turn and start moving back in the direction of the hut. As we walk, I decide to pose the same question to Enid that I asked Sue Sayer. 'What's the fascination of grey seals for you?'

'Gosh, I don't really know,' she says, then after a few minutes' reflection, speaks of the addiction of ID-ing, of spying a seal that hasn't been recorded before. 'It's like trainspotting.'

Back at the hut, a plan's being hatched among the volunteers to go home for lunch, then reconvene here in the afternoon. Terry and Alec make a show of acting like this schedule, this obligation to return, is profoundly onerous, yet it's clear that they love every minute. As Enid observed, the urge to 'collect' different seals as twitchers collect bird sightings and philatelists stamps has become a delicious obsession – one that also happens to be very valuable in terms of the information they're feeding through to Sue.

'Piece of advice for you,' says Alec with a pseudo-grumble as he slots his binoculars back in their case. 'Pack this seal business in before you start.'

I'm already travelling down the obsession highway myself and on my last day in Cornwall, 'this seal business' takes me back to the sanctuary at Gweek. Having remembered Dan Jarvis's exhortation to return in winter to learn about all the season's rescues, I begin my visit at the Seal Hospital where, back in September, only one pup, calling forlornly and shielded from human view, was receiving treatment.

Dan's smile, fringed by a stipple of stubble this time, widens when I ask how she's been doing in the intervening months. 'Really well – we were able to release her last week. She was off like a shot, zoomed out to sea pretty quickly. Good news!'

The pen that she was occupying now houses a very small, yet already moulted, pup. He's resting on a blue mat under a heat lamp, alternately yawning and chewing his front flipper. A gentle slope leads down from the mat into a small pool though he looks too lethargic to enter it. On the wall outside his pen is an info board written in upbeat sanctuary-speak – 'I am learning how to swim and eat fish for the first time! I will go down to join the other pups outside soon!'

I've never seen a mottled pelt on such a tiny pup – by the time they start losing their white birth fur, healthy pups are always such barrels of blubber. I learn, though, that he was extremely malnourished, having got separated from his mother in the days before he was rescued, and while the formula feeds he's been receiving here in the hospital are nutritious, their calorific content can't match the milk of grey seal cows. 'We're trying to get him onto fish now so he'll put on more weight – mackerel's good as it's quite fatty,' Dan explains. 'He's being force-fed at the moment but that won't last long. We'll be able to persuade him to play with, then pick up, fish from the floor pretty quickly.'

In pen number two, another skinny, already moulted pup has decided to take a tentative swim around her pool. At first she dunks her head under the water while the rest of her body remains in contact with the mat, then gradually allows her whole self to slip in.

'When the pups are at this stage of their rehab, it's a good chance for them to get their swimming action and buoyancy right,' says Dan. 'Often, there'll be bums in the air, flippers going in all directions. It can be quite hilarious.'

With no mother to coax and coach her, as I've sometimes seen happen at Aber Felin, she makes just a few uncertain circuits of her pool before hauling herself out onto the mat again, while I haul myself out of the low light of the Seal Hospital, to emerge, blinking, into the winter sunshine, ready to see some pups who are further along the road to recovery.

'We're pretty much at capacity at the moment,' Dan reveals as we walk downhill towards the Nursery Pools. After a quiet autumn with settled weather and only a few rescued pups in residence, 'December went a bit crazy. We got loads of calls with loads of pups reported to us in a very short amount of time. It's because of these storms that keep rolling through. The weather pattern's definitely changing – climate change is one of our biggest concerns. It's going to have a big impact on the way we work here at the sanctuary in the future.'

I bypassed the four Nursery Pools on my first visit to Gweek as they were empty of both water and seals. Now, each contains water that's deep enough for learning to dive plus several pups, all of whom look plumper and more robust than those in the hospital. A move into these outdoor pools is contingent on the pups being clear of all infections, and at least fifteen kilos in weight, with a blubber layer that negates the risk of hypothermia.

The moment that Dan and I appear, three pups break off their underwater chasing, bobbing up to the surface, their dark eyes riveted on us. A fourth, by contrast, snoozes on the poolside, his black-and-grey pelt streaked with the blue antiseptic spray with which some wounds were treated when he was rescued.

'Here, the pups get grouped up for the first time,' Dan tells me. 'They start to socialise and compete for food. At this point, we pretty much hide from them and throw the fish over the walls into the pools so they don't see where it comes from.'

Though there are few other visitors to the sanctuary today, there's still a loudspeaker announcement about feeding time being imminent at the Residents' Pool, which Ray, Snoopy and co. are currently obliged to share with a rabble of thirty-three pups who are in the final stage of recuperation.

The policy of trying to ensure that the provision of fish isn't associated with the presence of humans has met a bit of a hitch here. All the pups have learnt that a loudspeaker announcement means dinner's due as they're all jostling and hustling in the water near the gate where the animal care workers enter to feed the blind residents. I can make out Ray's larger head in the midst of all the little, dappled pup heads and though he's being bumped and barged and caught in the crossfire of lunges and snarls, he seems very tolerant. Snoopy, as usual, is keeping her distance, avoiding all contact with the hungry pups, waiting quietly on the concrete.

As soon as the animal care workers start tossing fish over the side of the surrounding fence and from the mezzanine viewing level above, there's a maelstrom of waves, spray, swashing and splashing as the pups surge up and down the pool, nipping at, and hurdling over, each other as they try to follow each fish's trajectory and anticipate where it'll land. Once a fish is grabbed, some pups adroitly flip it lengthways into their mouths and swallow it head first, while others hold it horizontally like a dog with a stick, then rip it into pieces with teeth and flipper claws. A mob of herring gulls has also swooped in, skimming over the heads of the pups, pouncing on any fish that's stranded on the edge of the pool, telegraphing their intention with a cackle of possession.

'How can you be sure they're all getting the amount they need?' I ask as a dark-pelted pup bursts whiskers-first through the surface of the water and snatches half a fish from another's jaws.

'Well, it's basically a big free-for-all by this stage,' Dan admits. 'If we see a pup struggling, we can move it back to the Nursery Pools but we haven't had to do that this year.'

A pair of animal care workers are inside the pool area now. One taps a pole along the concrete, coaxing blind Snoopy to follow, keeping her as far away from the scrimmage for fish as possible. Ray, answering to his name, also clambers out onto the side so that he can be spot-fed his quota. Several savvy pups try to hoist themselves out in the hope of being hand-fed too but most are still focused on the mid-pool tussle. Every time Dan or I speak, we have to raise our voices to be heard above all the diving and gurgling and snorting.

Once the final fish has been consumed, the waves that have been created progressively abate. The pups cruise through the water at a less frenzied pace now, double-checking that they haven't missed any morsels of fish and gradually turning their attention to other diversions. A bucket of kelp is dumped into the pool as an enrichment activity, provoking a number of the pups to start tugging on individual strands, and mouthing them to see if they're edible. A couple of other pups scramble onto the pallet that's floating in the pool, hitching a ride to the other end.

As I continue watching, I'm joined by a trickle of visitors. A woman in a micro-skirt, her legs tattooed with cold, reads the life story of Snoopy on one of the info boards attached to the fence. 'That is *so sad*,' she says, though her sadness is straight away supplanted by 'Are there any walruses here?'

Having eaten well, several of the pups are starting to be over-come by the urge to sleep, heaving themselves out onto the poolside and preserving the space around them by flapping their front flippers against their bellies whenever another pup gets too close. It's an endearing, mini-version of the adult behaviour I observe at Aber Felin.

'Quite a few of them are ready to go now – thirty-five kilos is the target weight,' Dan tells me. 'We release them in groups of two to seven on the coast where they were rescued. They're all flipper-tagged and numbered, as you can see – a blue tag in the left rear flipper for males and the right for females. Over the years, we've had re-sightings from as far away as the south coast of Ireland, County Kerry, the Bay of Biscay. . .'

Directly in front of me, flipper-tagged pup number seventy-four yawns with eyes squeezed shut and gives her pale grey head a scratch. It's extraordinary to imagine the sea journeying that could lie ahead so soon after her poorly days as a hospitalised pup. As she learns to transition from eating dead fish that rain down regularly from above to locating and catching live fish underwater, it's possible that she'll travel immense distances. Dan mentions a pup who was satellite tagged in North Wales at the age of three weeks and thereafter swam to Rosslare in Ireland, the Isles of Scilly and the north coast of France, finally winding up on the Lizard Peninsula just nine weeks later. 'In that time, she travelled a thousand kilo-metres – and that's just straight line distances, not taking into account the diving!'

'Incredible.'

'Yep. Seals have many mysteries.'

His comment echoes Sue Sayer's warning about the challenges of fathoming the behaviour of seals. While I've acquired so much more insight into the rhythm of, and interruptions to, their lives during this return visit to Cornwall, I realise that, like a suckling pup gaining two kilos daily, the weight of all that remains to be known has also increased enormously.[2]

4

Dye Hard

The bull seal is sleeping, body submerged, with just his face and the concertina skin of his neck exposed above the sea's surface. Every so often, his respiratory rhythm changes, nostrils flaring with a series of inhalations, then closing as he settles into another few minutes of held breath.

It's a breathing pattern consistent with slow-wave sleep, the deepest and most restorative stage. Usually, I'd relish the chance to watch a top predator at his most unguarded, but that's not the case today. The fact that I don't need binoculars to be aware of his most intimate breathing variations means we're too close. Much too close.

I, and my chocolate Labrador, Hooper, have joined a seal-spotting summer boat trip from the South Pembrokeshire resort town of Tenby. We're squeezed between a woman in an extravagantly padded jacket and a man who lost half his pasty to the swoop of a herring gull while we queued on the quay to get on board. He's been cursing the gull's gall ever since, his grumbling a counterpoint to the chug of the engine as we made for Caldey Island a few miles offshore. His reaction set me pondering on how graciously and sensitively we'll navigate our ever-expanding leisure spaces alongside gulls and other creatures, seals included, in the future.

As we approached Caldey and the guano-streaked cliffs of tiny St Margaret's Island off its western tip, I spotted the cruciform of a cormorant perched on a rock while another four skimmed the surface of the water.

'Bloody seagulls,' said my neighbour.

Best known for its Cistercian monastery and the Abbot's Kitchen chocolate bars produced by the resident monks, Caldey is also said to offer plenty of seal-sighting possibilities. Having glided alongside a stretch of white sand, we've reached a low cliff where, our guide has informed us, a young seal is swimming to the left of the sleeping bull, triggering a rush to the side of the boat and a chorus of 'Aws' from my fellow passengers.

Instead of moving on to see if more seals can be found elsewhere, the decision is taken to milk this sighting for all it's worth and our boat cruises back and forth, back and forth, so that people sitting on both sides have maximum viewing opportunity. And each time the boat turns, we close in on both bull and juvenile a little more.

I'm aware that there's a Pembrokeshire Marine Code – doesn't the Code of Conduct for Seals recommend keeping fifty metres away? – and our current proximity doesn't feel at all comfortable. The bull isn't bothered at the moment – he's still deeply asleep, his nostrils still flaring at irregular intervals whenever he needs to take in more air. But if he were to wake, wouldn't he be startled by the boat's looming presence? And to avoid getting boxed in against the cliff, the juvenile's left with no choice but to dive and shoot away.

Back and forth, back and forth again. Over the boat's loud-speaker, the guide is relaying information about the 'sixty to eighty grey seals in the Caldey Island colony' but no one is paying much attention. In fact, most people have already grown bored of watching the solitary sleeping bull too and a number have shifted their attention to Hooper, who's far more alert and responsive. Throughout the rest of the trip, there's continually someone stroking him, fondling his ears, encouraging him to plant his front paws on their knees and engaging me in dog chat. The only time the

Hooper focus gets diluted is when another tour boat passes close to ours as we scud back to Tenby and everyone on board both of them feels compelled to wave.

By the time we dock in the harbour beneath its colourful tiers of Georgian houses, wavelets of sadness are ruffling my surface. Sadness for the seals who have to withstand the demands of our desire to see them and the withering of our interest the instant it's fulfilled.

Sadness, too, because having encountered seal disturbance issues on land near St Ives, I've now, some eight months on from my winter visit to Cornwall, had my first introduction to disruption at sea.

Several days later, I've swapped the cramped bench on the boat for a seat at the blue Formica kitchen table of my childhood. Dad, with Hooper at his feet, is sitting opposite, polishing off a bulging baguette – even though he's now in his eighties, I'm grateful that he's still in robust physical health with just a mild heart murmur that gets checked once a year. He remains intensely independent too, content to live on his own and care for himself. My Pembrokeshire home is over three hours away but I still manage to see him regularly and we've spent many happy days in the years since my mother died walking, bird-watching and sharing favourite nature-themed books and poetry. Today, as well as bringing along some lunch to enjoy, I need to reclaim a book that I lent him a few months ago, one that relates to the next phase of my seal journey.

'Nice sandwich,' he says, brushing crumbs from his red fleece with the word 'Intrepid' branded above the heart.

'D'you want some fruit?' I indicate the selection I've piled into a bowl, a pair of bananas hauled out on the top like seals curved over a rock.

'Yes, I'll have some of the – um . . .' Pause. Slight frown. 'The – um – melon, please.'

'Mango,' I correct as Hooper hoovers up the baguette crumbs. 'I'll cut it in half – we'll share it. I'll put the kettle on too.'

'Right-o. I'll have a look for your book while you're doing all that, then. Hardback, you say?'

'Yes, with a stormy sea and an island on the cover.'

For the next few months, with the pupping season having just got underway for another year, I'm going to be focusing on grey seals in Wales and will initially visit Skomer Island off Pembrokeshire's far west coast. Skomer's one of the most important seal sites in the UK – studies got underway in 1974 and, since 1983, a succession of wardens have taken a more systematic approach, producing detailed annual pupping reports, most recently commissioned by the Welsh government-sponsored body, Natural Resources Wales. Number of births, survival rates up to the first moult, causes of mortality, and instances of disturbance and pollution, including entanglements, are all documented. My pre-visit prep has included accessing and reading these reports online, arranging to meet the island's current wardens to find out more about their pup monitoring work and reconnecting with several books in which Skomer seals play starring roles.

Waterfalls of Stars, the book for which Dad is hunting, is Rosanne Alexander's account of her ten years on the island from the heatwave summer of 1976 onwards, and it offers a compelling insight into both Skomer's outer, and her own inner, topography. She was aged only twenty when she and Mike, her husband of just a few days – their marriage was, astonishingly, a condition of his appointment as warden – arrived on the island. The book recounts countless challenges – trying to eke out dwindling food supplies when persistent treacherous winter weather precludes visits to the mainland, adapting to a life without electricity, kitchen windows shattered by Atlantic storms – yet she also writes rapturously of her deepening connection with every cave, raven, rock and razorbill and her increasing disinclination to leave them, even for a day.

Seals make only sporadic appearances in the first quarter of the book, but thereafter, they gloriously dominate and she's privileged to be privy to untold intimate moments of their lives. She witnesses the delight of a pup's first attempt at swimming and the devastation of a breeding beach smothered by oil. She watches seals glistening in moonlight and a cow lying next to her stillborn pup for day after day, periodically enclosing it with her foreflippers. Having to wrench herself away from Skomer after a decade following Mike's acceptance of a conservation job elsewhere has such a seismic effect on her that even now, more than thirty years on, she's still not able to bring herself to return.

Though *Waterfalls of Stars* wasn't published until 2017, Rosanne previously wrote many of her Skomer experiences into a novel, *Selkie*, that appeared in the early 1990s. In many ways, the seal pup hero is a Rosanne stand-in, with his love for the island and reluctance to leave merging with hers. At times, she even seems to forget that she's writing from the pup's perspective rather than from her own, with the narrator describing aspects of Skomer, such as the inland flowering of bluebells, about which Selkie can have no knowledge. Almost half the book is devoted to the first three weeks of Selkie's life, the period with which Rosanne would be most familiar, while the chapters in which he leaves Skomer and swims north, are nowhere near as vivid. He survives a series of life-threatening incidents – an autumn storm, a boat propeller injury, an oil spill – and, as always when I embark on an animal narrative, I proceeded slightly queasily the first time I read it lest a final chapter death be on the cards. My fears were unfounded though – Rosanne has far too much affection for seals to allow such a tragedy to befall the eponymous hero of her novel.

'Is this the book you're after?' Dad's ambled back into the kitchen with *Waterfalls of Stars* in his hand and Hooper at his heels.

'Yep, that's the one.'

'Looks terrific.' He runs his fingers over the turbulent sea on the front cover, opens it, peruses the first few pages. 'Can you bring it back again when you've finished with it? I'd like to read it.'

'You've already read it, Dad,' I remind him. 'About a month ago.'

He shakes his head. 'Not this one.'

'Yes, we had a chat about it.'

He scrutinises the cover once more, reopens it at a random page, tilts his chin so he's looking through his bifocals at the right angle and shakes his head again. 'Nope,' he says with a tinge of irritation. 'I've definitely never read this.'

Unease needles the length of my spine. I distinctly remember talking to Dad about his response to reading the book. He said that even though he enjoys his own company, learning about Rosanne's experiences made him realise he wouldn't like to live in such an isolated location, but he'd still quite fancy a Skomer day trip the next time he visits me in Pembrokeshire. I've grown used, over the past few months, to subtly helping him out when he loses the odd word – 'mango' seems to have permanently disappeared and 'melon' is always what he thinks he's looking for – but this is something new, something more significant. Is it a one-off? Will he recall having read the book tomorrow or at some later date? What else might disappear into the gulch of lost memories? And how critical is this forgetting?

He sits down and pushes the book towards me across the blue Formica, a gesture that says he wants to draw a line under our exchange. 'Now, what about that cup of tea?'

On the map, Skomer looks like a dividing amoeba, comprising a larger western blob of land, a smaller eastern, and an isthmus like a stretch of protoplasm connecting the two. I usually visit in spring or early summer when the island's a colourful flourish of breeding puffins, bluebells and red campions, and hundreds of other hopeful islophiles vie for the limited number of daily landing tickets. Today, though, waiting to board the boat in the little cove of Martin's Haven in early September, there's just a group of primary school-children, a cheery young teacher, an older, jaded-looking one, and a woman who's hankering after Cornwall rather than appreciating

West Wales. 'Are you in the queue for Ross Poldark?' she asks me. 'I can just imagine him galloping over those cliffs!'

Jack Sound, the half-mile body of water separating the mainland from Skomer, has the reputation of being a duplicitous strait with concealed reefs and fierce currents. But this morning's crossing isn't at all hazardous – we're free of wind and big waves and, in contrast to springtime visits, largely free of seabirds too. In the absence of ornithological distractions, one of the schoolboys plagues the older teacher with a series of burning questions – 'Sir! Sir! If there was a fight between a great white shark and a killer whale, who'd win? . . . Sir! Sir! If I was to jump off that cliff, would I be alive or dead?'

The Poldark fan runs a few questions past me too. 'Will there be much to see when we get there?'

'Yes, there should be plenty of wildlife,' I tell her.

'Oh, wildlife, lovely,' she says, then lowering her voice, leans in closer, 'What about toilets?'

On arrival, once we've all slogged up the long, steep flight of steps from the North Haven boat jetty to the cliff top, an employee of the Wildlife Trust of South and West Wales, which manages the island, delivers his 'Welcome to Skomer' speech. Amidst all the advice about the importance of keeping to the marked footpaths and the location of the compost loos, he shares some seal pup information – thirty-eight have been born in the island's caves and coves so far this season, with around 180 expected in total.

Being familiar with both the annual Skomer Seal Reports and the island's network of paths, I decide that my best pup-spotting bet will be to head for a viewpoint overlooking the eastern section of the splitting amoeba, known as The Neck. As I walk, I'm aware that though there's no trace of the shrieking kittiwakes and huddles of guillemots that crowd the cliffs in spring and summer, there could be a few hundred thousand young Manx shearwaters in burrows beneath my feet. Even without the occasional 'CAUTION BURROWS' warning chalked on a stone next to a hollow in the middle of the path, it's impossible to forget that Skomer's riddled with holes like a giant Emmental. The island supports one of the most significant breeding populations of Manx shearwaters in the world and at this stage of their cycle, the adults have left to migrate to their wintering grounds in South America. Their fattened-up chicks remain for a while longer, emerging from their burrows mostly at night to stretch their wings and, it's believed, to 'stargaze', building up a mental constellation map that will help them navigate and orientate themselves in the future. Very soon, they, too, will embark on an extraordinary journey and complete the six- to seven-thousand-mile flight from Skomer to the coasts of Brazil and Argentina without adult guidance in as little as two weeks.

In as little as ten minutes, I reach my planned seal pup viewing point, where a seasonal Skomer volunteer has set up a telescope, positioned at a low level in preparation for the schoolkids. She's reluctant to readjust it so I stoop and peer through: it's trained on Driftwood Bay, one of the island's busiest pupping sites. At the cliffs' base are several pebbly indentations which will enlarge and amalgamate at low tide into a single shingle beach. Now, though, the high tide is forcing three newborn pups up against the cliff face while the pied heads of their mothers are visible in the water, keeping a fretful watch. A plumper, sturdier pup is rolling around on the pebbles, leaving streaks of green algae on its white coat. It's a week or so older than the others, with front flippers and face that are darker than the rest of its body, indicating that it's starting to moult.

As soon as the school group arrives, the volunteer hustles me away from the telescope and tells the children to form an orderly queue.

'You can have a little snack while you're waiting,' says the enthusiastic young teacher. '*Not your lunch, Cameron!* A snack. Good old Mr Collins has brought some biscuits.'

While one child after another takes a turn at the telescope, I swap to binoculars. A red mark on the flanks of each of the young pups has caught my attention. At first I think they must have all been wounded by being thrown against the rocks by the waves but then I grasp that no, it's actually red dye. Here on Skomer, the wardens who monitor the pups spray the newborns as this makes it easier to keep track of births and know for certain if there have been any new arrivals since the previous survey. It strikes me as quite a controversial monitoring method, though, and one which is now out of favour with many other seal sites. Just across Jack Sound on the mainland, dye-marking was phased out over a decade ago: designated marine conservation officers would access the coves by boat, then swim to the beaches to spray the pups but since the cows would sometimes flee into the sea and were on permanent high alert, the decision was taken to eliminate the disturbance.

It looks like the Poldark fan is on high alert right now, placing her very white sneakers very precisely between rabbit shit and the feather-and-bone remains of Manx shearwaters preyed on by great black-backed gulls. As she waits for her turn at the telescope, I carry on training my binoculars on Driftwood Bay. Unexpectedly, one of the rocks resolves itself into the shape of a seal, this time a fully moulted pup with a mottled, adult-type coat rather than white birth fur, consummately camouflaged against the pebbles. Right on cue, good old Mr Collins launches into an explanation about how camouflage has evolved over thousands of years to enable seals to blend in with their surroundings when they're on land, making it difficult for humans, and other potential predators, to see them.

'Sir! Sir!' The pupil with all the questions again.

'Yep?'

'Can seals see other seals easily?'

For the next few hours, I walk the four-mile circuit of the island, looking out for more seal pups and enjoying encounters with a variety of birds. Pre-migration swallows switchbacking just above my head. A tumbling flock of choughs at The Wick, a vertical cliff fronting a thin inlet. A coven of ravens on the rocks at Skomer Head, their tail feathers brushing yellow lichen.

I enjoy views of Skomer's sister island, the plateau of Skokholm, too. There, in the 1930s, the Welsh naturalist, Ronald Lockley, established Britain's first bird observatory. He also pursued pioneering seal research in Pembrokeshire in the 1940s and I have every expectation that I'll cross paths with his work and legacy again in the months to come.

For the time being, though, I intend to cross paths with one or both of the Skomer wardens.

'The Seal Woman's here!'

I've hopped over the chain that prevents members of the public from wandering along the track leading to the Island Office, passed a cluttered workshop containing a knackered tractor, and arrived at the door of the wooden, solar-panelled wardens' house. Joint warden Ed Stubbings has answered my knock and announced my arrival by calling down the hall. Though I'm absurdly pleased to be referred to as the Seal Woman, his flustered tone is a bit perturbing: it turns out that he already has a meeting underway with a group of volunteers so won't be able to chat to me. Happily, however, I'm still able to speak with his wife and co-warden, Birgitta Büche, otherwise known as Bee.

Leaving behind the animated voices and appetising soup smell wafting from the kitchen, I follow Ed's directions upstairs to the library, a bright and airy room with views on three sides. The house is located on the isthmus connecting the western part of Skomer with The Neck and when Bee shows up – freshly show-ered, rosy cheeked, nose stud glinting – she finds me at one of the windows, craning and straining to see if there are any seals on any of the beaches below. Originally from Konstanz on the German-Swiss border, she job-shares with Ed both the Skomer warden's role and the contract to undertake seal research and produce the annual report. 'It's a very important study as it's been going on for so long,' she tells me, claiming that 'There's nowhere else in Britain where seals are being looked at so intently.'

Though I'm itching to dive in with a question about dye-marking,

I start with the less controversial request to hear what she finds most compelling about grey seals.

'Well, first of all,' she says, 'the females are amazing. Some come to the same place on the same beach each year and give birth on the same day. Others go one year here, one year in Cornwall and then disappear for a few breeding seasons before coming back here again. Usually a species is site faithful or it's not – it's extraordinary to have such variation within one species.' Sinking onto a chair, she kicks off her Crocs and curls her stripy-socked feet beneath her. 'The pups are amazing too. Our beaches often get flooded at high tide so the pups have to swim from day one. I've watched a whitecoat, just a few hours old, swimming like a water baby. And when it got tired its mum swam underneath so it could get on her back.'

Impressed both by Bee's fulsome answer to my question and her ability to express it in faultless English – were it not for her residual German accent, I'd never guess she wasn't conversing in her first language – I mention the high tide I witnessed at Driftwood Bay earlier. It looked like the new pups were going to be forced to swim then too, though it's popularly accepted that they can't do so at such a young age because their white birth fur isn't waterproof, and unless they've moulted and acquired their adult-type coat, they'll drown. Yet while it's fascinating that the pups here are, by necessity, often disproving received wisdom, there are still presumably instances when they get into difficulties and suffer injuries, so I ask Bee about the Skomer approach to carrying out rescues.

'We don't do rescues,' she says, though qualifies her comment by adding that she and Ed once cut a juvenile free from fishing net and it's imperative to intervene if a problem is human induced in such a way. She also refers to a colleague who believes there's a tendency towards over-intervention – 'you may do it just to make your own heart feel better. For example, if a pup's been abandoned by its mum, there could be a good reason for that. It might be unwell and she knows it won't survive.'

I can see the logic in opting to orchestrate a rescue if a pup is suffering from an anthropogenic injury like a fishing-gear entanglement but otherwise choosing to let nature take its course, yet it still feels counter-intuitive not to always try to alleviate the suffering of a sentient young creature. It makes for a marked contrast with the approach favoured by Sue Sayer and the Cornish Seal Sanctuary too.

Fuelled by a mug of tea supplied by Ed, I decide it's time to broach the subject of the Skomer pup monitoring process.

'We walk round the island every day, record all the newborns and look at how the older pups are getting on,' Bee reveals. 'And we dye-mark no more than once a week to keep disturbance to a minimum. We watch from the top of the cliff and if any pups are suckling, we wait fifteen minutes till they're done. Then we abseil or scramble down to the beaches and caves – a lot of our pups are born in caves and spend the first few weeks of their lives in complete darkness.'

Though I don't shy away from suggesting that spraying the newborn pups – albeit to give the most accurate assessment of numbers and continue Skomer's long-term research programme – must still cause disturbance, Bee insists that things have changed in terms of how bothered 'the mums' are by human presence. 'One of the former wardens, Mike Alexander, came back to visit and got permission to scramble down to Driftwood Bay and afterwards he was nearly in tears. He said it was one of the nicest things he's ever done. When he was the warden here he'd never see the mums on the beaches with their pups – they were all so scared, they'd go into the water as soon as they got the slightest glimpse of someone on the cliff. And when they came onto the beaches to suckle, they'd do it as quickly as possible and go straight back in the sea.'

I recall reading about this behaviour in *Waterfalls of Stars* and how Rosanne Alexander always needed to do her cliff-top observing from a hide. In the book's final few pages, however, she writes about having heard that the Skomer seals are not so nervous now and that lack of fear is allowing mothers to spend more time with their pups. Though she used to wonder if the 'restless anxiety' of Skomer seals was due to competing for limited space on safe beaches, pup numbers are now three times what she used to see and the seals are at ease with each other and 'barely troubled by people'. With changing attitudes and better protection, she optimistically suggests that seals are 'finally forgetting what it is to be persecuted'.

'Yes, that's exactly what Mike thinks,' confirms Bee. 'Because seals are such long-lived animals – in the wild, females can live for thirty-five years and males for twenty-five – he thinks there are some out there that are the young ones of those who could remember getting shot. They will have been taught to be scared of humans but gradually that gets thinned out and relaxed mums start to give birth to relaxed pups.'

In my wider, pre-visit readings about Skomer, I came across several references to nineteenth- and twentieth-century seal shootings, both for sport and because fishermen viewed them as competition,[1] so feel gladdened that there seem to be signs that Skomer seals' collective memory of persecution might be fading. I'm gladdened, too, by the anecdotes of quirky seal encounters that Bee proceeds to share with me before I need to leave for the mainland.

'We've had lots of immatures in our Zodiac, our inflatable boat. It's funny to watch how they get in – they jump out of the water and bounce against the boat and if they don't quite make it they bounce off again.' On another occasion, an adult bull seal who managed to haul himself into the Zodiac showed little inclination to ease himself back into the sea. 'We thought he'd hop out when we rowed up to the boat as bulls usually keep their distance but he just lay there and looked at us, and giving the boat a slap didn't work either. So we had to wait for a few days to use the Zodiac until the bull was gone.'[2]

I'd love to stay and listen further but the last boat of the day is due to depart. I bid Bee goodbye, hare along the track from the house and down the steps to the harbour, then tumble into the boat with the Poldark fan, schoolkids and teachers. As we zip back towards Martin's Haven, I reflect that, though I still have reservations about dye-marking, Bee and Ed's decision to postpone their use of the Zodiac to accommodate the presence of the bull seal is positive indeed. In contrast to the man whingeing incessantly about the gull stealing his pasty in Tenby harbour, it's a small but admirable example of the consideration and flexibility that are required if we're to successfully live side by side with another species.

5

ExtractCompare

For the past hour, I've been shadowing a man – grey-bearded, tanned and rangy – who's carrying not a rucksack but a Tesco shopping bag. He's a fast walker and easily outpaces me but every so often, when he detours briefly off the rock-strewn coast path, I have the chance to catch up with him again. And each time I do, his bag is bulging a bit more with the wild autumn mushrooms he's gathering. He's deeply focused on this foraging, and doesn't bother to acknowledge either me or the silhouetted view of Ramsey Island, the next destination on my seal journey, half a mile or so away across the sound. He pays no attention to the gush of the tidal race rushing between island and mainland or the posse of harbour porpoises whose broad dorsal fins rise then slide away, rise then slide away, like the details of a dream that's refusing to be remembered. And when I stop near the crumbling edge of a pebbled micro-bay, he strides on, not in the slightest bit interested in looking for grey seal pups either.

The North Pembrokeshire cliffs are uncharacteristically low just here and I make myself low too, inching along on hands and knees and doing a slow motion lean over the edge to minimise disturbance. My stealth is rewarded by the sight of an almost-moulted pup suckling vigorously from its mother, then rolling away and falling asleep on its back, no more than ten metres below me. The cow likewise seems set on sleeping, but her pre-nap scratch is interrupted by a huge black bull seal, jaws amorously agape, launching himself from the shallows towards her. Thanks to his keen sense of smell, he's perhaps detected that she's coming into season again but she's not yet willing to tolerate his advances and wards him off with a howl and a flap of her front flipper. I expect him to persist but instead, he heeds her warning, gives a hissy exhalation and retreats towards the water. Though the cow's intent on settling down to sleep again, she remains on semi-alert, opening her eyes at intervals and raising her head to check that the bull's not about to make a fresh approach.

For the time being, it appears he has another cow in his sights. In the froth of waves just offshore, a wholly moulted pup, the pale background colour of her pelt indicating that she's probably a female, is playing with her mother. They periodically touch noses, then roll together, twisting and twining and snaking round each other as if performing an underwater salsa. It's an enchanting scene, and a poignant one, as it can only be a matter of days before the cow heads out to sea, leaving her weaned pup to fend for herself, with no further contact between them. After typically undergoing a post-weaning fast, during which time she may lose up to a quarter of the body mass she's gained while suckling, the pup must work to perfect her diving technique, and locate and catch randomly distributed prey, all without the benefit of learning from the foraging skills of her mother. Critically, she must do this within the thirty-six or so days it takes before her remaining blubber and protein reserves are depleted and starvation sets in.

In view of the challenges she'll have to face, I can't help but feel irritated when the bull torpedoes himself towards her mother and brings their play to a premature end. Spooked by the bull's bulk, the moulted pup swims, quick as a fish, towards the beach, then hefts herself across the pebbles to a place of safety.

As I watch her nose a ribbon of kelp in the tideline with dog-like curiosity, a coast path walker tramps up and parks himself a few metres away from me. By the time he's finished rummaging in his rucksack for his phone, had an extravagant sneeze, ostentatiously munched a packet of crisps and turned his baseball cap back to front so the peak doesn't interfere with his binoculars, all of the seals, with the exception of the deeply sleeping moulter, are alert to his presence and, by default, to mine too. The playful pup has abandoned her investigation of the kelp and scooted further towards the back of the beach while the semi-dozing cow now has one eye on possible bull activity and one on our human doings on the cliff top.

Hoping the man will follow my lead and leave them in peace again, I discreetly withdraw from the cliff edge and walk on. Over the course of the next hour, I see no more pups – not even in the cove unofficially known as Seal Bay, one of several thus named in Pembrokeshire – but I hear plenty. From inside the sea caves, over the top of which the coast path passes, hunger wails are echoing. It's not in the least bit surprising that the lifeboat's been known to launch from the nearby harbour of St Justinian's in response to a call that a child could be heard crying, believed to be stranded somewhere at the base of the cliffs, plainly in distress.

It's from St Justinian's that I travel to the RSPB reserve of Ramsey Island, about eight miles north of Skomer as the chough flies, the next day. The mini-ferry makes quite leisurely progress, but I and its only other passenger, a taciturn man with multiple camera bags, are still invited to move away from the seats at the stern to avoid getting drenched by water crashing over the side. A chill easterly wind is blowing, triggering atypical wave action, quite a contrast to Pembrokeshire's usual warmer southwesterly. The skipper, appealingly piratical with his hoop earrings and red spotted head-scarf, also invites us to check our rucksacks for rats: they were eradicated from the island in 1999–2000 in a programme aimed at protecting nesting seabirds, and care continues to be taken to ensure that visitors don't unwittingly reintroduce them.

As soon as I clamber off the boat, my attention is hijacked by the portliest white-coated pup I've ever seen lolling on the pebbles not far from our docking spot.

'Twenty-three days old – and he's only just starting to moult. His mother's still around too. So unusual!'

The woman who's joined me on the walkway overlooking the little beach must be Lisa Morgan, Ramsey's co-warden, with whom I've arranged to speak today before doing some exploring. With a walkie-talkie clipped to her waist and a scarf preventing her long reddish hair from being flung around in the wind, she has startlingly pale eyelashes and eyebrows and a welcoming smile. I warm to her instantly, delighting in the fact that she's introducing me to one of the pups she's monitoring before properly introducing herself.

I'm delighted, too, that she's so effortlessly able to identify this pup as 'he'. Sexing whitecoats is tricky – at this pre-moult stage, the only way of knowing is by checking the lower abdomen for the presence or absence of a penile opening, midway between the umbilical scar and hind flippers. Newborns' folds of skin make it awkward to get a glimpse but this older pup, helpfully lying on his back and flaunting his entire underside, has gained so much weight that his skin is as taut as cling film over pastry dough and the small penile hole is clearly visible.

I follow Lisa uphill to her whitewashed farmhouse home and, though it's cold, we choose to sit on a bench outside. She confesses that she's wearing thermals for the first time this season and though I'm not as well prepared for the sudden temperature drop, I'm keen to enjoy the view across the seethe and lather of Ramsey Sound. I'm keen to make friends with Dewi the Border collie too: while Lisa disappears to make tea, he conducts a thorough examination of every part of me that his nose can reach, familiarising himself with all the smells that I've brought with me from the mainland.

By the time she returns, he's moved on to forensically sniffing every square inch of my rucksack. 'You've got to remember he doesn't get out much!' she says, adding that he's a working dog who's lived on Ramsey since he was ten weeks old. Lisa and her husband had to learn to farm when they took on the job of co-wardens of the island well over a decade ago: there's both a flock of around two hundred sheep and a small herd of red deer to manage.

I wrap my hands round my mug of tea to warm them and listen to Lisa expand on other aspects of her work – most notably, the contrast between the Ramsey seal surveying process and that on nearby Skomer.

'Natural Resources Wales don't fund any of our seal survey work – it's part of our RSPB job – whereas Bee and Ed get paid for the whole research project. We do it mostly in our own time too – we don't have time to survey as part of our normal working day. That's fine, though, because I love doing it and it was me who got it written into our management plan in the first place.'

'It wasn't part of the original job description?'

'No, and it wasn't that the RSPB was guilty of focusing just on birds because there was lots of other stuff in the management plan – plants and lichens and everything. Just no seals. It was unbelievable! Ramsey's so important – over six hundred pups are born here each year – and it should be right up there in the literature when you read about seals.'

'How often do you carry out your pup surveys?' I ask as Dewi surveys the grassy slope beyond our bench, finds a stick, drops it at my feet, then backs off and waits with intense focus for me to throw it.

'Every third day all through the season. It takes me about six hours to do the survey and another six to go through all the photos.' She adds that she uses 'very clever' software for ID-ing the females. 'I'll show it to you later.'

'Do you dye-mark the pups?'

'No, we focus on nine beaches that can be seen from the cliff top and don't access them at all. It's done observationally with no interference, which I much prefer. We've worked hard over the years getting boats and kayakers not to disturb the seals so I like to set an example and do the same.'

I'm heartened to hear about Lisa's anti-disturbance stance. Struck, too, by how it's not just warden partiality but also availability of time and funding that's determining how pup monitoring is being undertaken on Ramsey and Skomer. While Lisa manages to fit in a pup count every three days alongside all her other wardening responsibilities, Bee and Ed, who are sharing a funded research role in addition to their work as co-wardens, can pup-count daily. From the seals' perspective of course, Ramsey and Skomer are part of the same wider pupping area: it's only we bureaucratic humans who have split it into separate zones, managed by different organisations and with contrasting monitoring methods.

While Dewi and I maintain our stick-throwing-and-retrieving rhythm, Lisa reveals her biggest monitoring bugbear – 'What I really hate is seal sites around the country collecting data and then being really possessive over it. It happens quite a lot' – then shifts to sharing her saddest Ramsey memory. 'There was a storm and on our biggest beach only four pups survived. I didn't get upset about that – we lose pups all the time – but the saddest thing was that all the cows came back. They travelled up and down the beach the whole day long, hunting for their pups, looking under boulders – it was frantic – you could see all their trails in the sand. It was the most hideous thing I've ever seen.'

Her story causes me to involuntarily shiver. In the months following Mum's death, I was drawn to reading about nonhuman animals' mourning rituals and expressions of grief – elephants repeatedly touching a deceased relative with their trunks, groups of magpies laying grass around another's corpse – and in the weeks since I became aware of Dad's faltering memory, my night-time thoughts have inadvertently hurtled back to this subject. The sleep-depleted child who was engulfed by irrational fears of parental death has made a comeback and now the image of the desperate cow seals seeking their pups has triggered another wave of anticipated loss and grieving.

'D'you need another cup of tea?' Lisa must have noticed my shiver and assumed it's the result of sitting too long in the wind.

'No, I'm fine, thanks.' Having planned to check out some of Ramsey's pupping beaches, I say I'll warm up by walking instead.

She explains that a marked route of almost four miles will take me to most of the notable viewing points and I pick up the *Llwybr/* Trail sign above the farmhouse, heading inland across an area of grazed pasture towards the west coast. Aside from the island's resident sheep, I glimpse the blue-grey back and head of a wheatear scuttling ahead of me, then the sunset splash of his chest as he takes flight.

While the clue's in the name – *mawr* means 'big' in Welsh – I'm still not prepared for the awe-inducing scale of the first bay I come to. Aber Mawr is Ramsey's most populous pupping beach, backed by the hulk of a cliff on which I'm now standing. I drop to a squat as close to the edge as I dare with the east wind so boisterous at my back and battle to steady my binoculars. Lisa told me there are seventy-two pups on the beach at the moment, though after several wind-buffeted recounts, I've still only spied fifty-nine. Then I smile at myself for thinking 'only' – a breeding colony, or rookery, containing fifty-nine pups is by far the largest I've seen.

I'm hoping I might be lucky enough to witness my first seal birth while I'm here. Lisa mentioned she's seen five on Aber Mawr this week and there are several very pregnant cows sprawled on the pebbles just below the tangled tideline of driftwood and kelp, orange floats and fishing net. I know some of the pre-birthing signs to look out for – a restlessness and circling, perhaps some digging with the front flippers – but I also know, from having watched so closely at Aber Felin, that there can be false alarms. Many a time I've been sure that a birth is imminent, only for the cow's spell of fidgety circling to give way to a three-hour nap and for darkness to arrive in advance of any pup.

In addition to all the cows and pups, I count eight bulls on the beach. Grey seals are polygynous, meaning that a single bull couples with multiple cows and a number of dominant males appear to monopolise the mating. It used to be thought that bulls established and defended territories into which females came to give birth but it's now understood that cows dictate the distribution of animals by selecting a preferred place in which to pup. A bull then takes up position within a loose group of cows and vigorously vies with other bulls to maintain his access to them. During this period ashore, he fasts, in some cases for over fifty days, and, like the nursing females, experiences substantial loss of weight.

Though bulls are sexually mature by the age of six, they're unlikely to be active breeders at that stage and certainly won't have gained beachmaster status, so the eight bulls below me on the beach are probably at least ten years old. I spot several smaller, younger bulls leaving the water, though, attempting to sneak up on any cow that's close to the sea's edge, much to the dominant bulls' displeasure. Unlike sea lions and fur seals, who are able to rotate their rear flippers beneath them and 'walk' at a rather faster pace than a human, the movement of grey seals on land can appear cumbersome – a lunging weight shift from chest to pelvis, front-flippered propulsion, a hump along. Yet, as the Aber Mawr bulls prove, they're still able to inject pace into their land-based activity, displaying, when impelled by fight or fear, unexpected and exhilarating speed.

Over the gusting wind, I can hear the bulls grunting as they face off and chase off, as well as youngsters yowling for milk, cows howling away lustful males, and pregnant seals hauling their unwieldy bodies over the pebbles. The beach is teeming with activity just as is described in *Halic*, a seal-themed novel by Ewan Clarkson that's mostly set on and around Ramsey, a copy of which I've brought with me in my rucksack for the pleasure of matching Clarkson's in-fiction descriptions to the reality of the bays I'm visiting.

If Rosanne Alexander's *Selkie* is the imagined life story of a young seal that unfolds against the backdrop of Skomer, *Halic* – whose name echoes the *Halichoerus* part of the grey seal's scientific name – is the Ramsey equivalent. Between the two novels, however, there are fundamental differences in approach. Every paragraph of *Selkie* is infused with Rosanne Alexander's love for Skomer, while in *Halic*, it is anger and polemic that are more often to the fore. The first years of Halic's life are beset with threats, from ghost gear to unexploded mines to toxic pollution due to pesticides washing into the sea. *Halic* was published in 1971 and it's dispiriting to think that it's only relatively recently that some of these threats, both to seals and the broader marine environment, have received wider attention and concern.

Halic adventures far south of Ramsey – around the Scilly Isles, Brittany, and even in the Bay of Biscay – but it's to his birth island that he returns when he reaches maturity, having dreamt 'uneasily of young maiden seals'. Seal sex is emphatically anthropomorphised, veering between the absurdly romantic and the soft porn of male fantasy. ('A pair of voluptuous virgins came up on either side of a lordly bull . . . They . . . caressed him with their strong young bodies, slid over his flanks with sleek, tremulous movements.') Clarkson lingers most lasciviously on Halic's mating with a young seal, Lugo, informing us that 'he took her, at first roughly and savagely, so that she moaned with the pain of his entry' but then proceeded to be more 'gentle and loving, so that all was fluid and light. The clean cold sea was their marriage couch and the cave their bedchamber.'

As a counter to the episodes of rapture and titillation is the novel's violent climax, a series of chapters foregrounding two fishermen who beat and kill a seal who's been consuming 'their' fish, and who subsequently sail to Ramsey to wreak greater revenge, with pickaxe and gun, on a whole pupping beach. Halic and Lugo both escape but their pup dies in the massacre. There is seal-on-seal violence in the novel too – Halic has a bloody battle with a beachmaster on the northeastern tip of Ramsey, for example – but Clarkson presents this behaviour as an instinctive, necessary, even noble part of the grey seal life cycle, in contrast to the premeditated and unjustifiable savagery perpetrated by humans.

I continue to keep one eye on the beach while riffling through *Halic* till my fingers are too chilled to properly turn the pages. Since none of the pregnant cows is showing signs of imminent birthing, I decide to stride on, leaving Aber Mawr and climbing steeply upwards.

The noise – loud, raspy, guttural – that greets me when I reach a heathery plateau is, on this occasion, non-seal in origin. A red deer stag stands, ankle-deep in dead bracken, and gives a proprietorial bellow in the direction of the nine gangly hinds that are grouped nearby, their pricked ears appearing way too large for their delicate faces. When they sense my presence, they start glancing twitchily between me and their surrounds as if assessing an escape route but the stag herds them together, tips back his antlered head and lets out another roar. The red deer is often a favourite reference point when an attempt is being made to describe a grey seal's size – 'Britain's largest land-breeding mammal, bigger even than the red deer' is the line I recall from Dad's animal encyclopaedia that I loved so much as a child – and I'd certainly concur that many of the bulls I've seen are bulkier than this stag. To my mind, though, it'd be far more interesting to compare the two species' autumn mating behaviours than body mass. If there were another stag around, he'd be bellowing and posturing, parallel walking beside, and possibly locking antlers with, his rival in a fight for control of the harem and the right to mate with as many fertile hinds as possible. I've seen grey seal equivalents of the parallel walk – two bulls sizing each other up, assessing whether to fight, expend essential energy and risk serious injury, or drop the challenge and retreat. Like the grey seal bull, dominant red deer stags fast and lose condition during this period too. As for differences in the species' mating rituals, the most glaring seems to be the behaviour of the females: compared with grey seal cows who snarl and snap and fend off lustful bulls until they decide, once in oestrus, that they're ready to mate, red deer hinds, with their prey animal preference for gathering in herds and no newborn fawns to protect, appear quite docile and compliant. I watch as they let themselves be shepherded through the heather by the stag, whose periodic roars then accompany my descent to the next bay and my re-entry into the realm of seal.

A cow scratching her pup with her flipper, hastening the moult by extracting tufts of white fur which the wind then whips away. A newborn's mewing echoed by the call of a buzzard overhead. From a cave, a cow's keep-away vocalisations, bizarrely reminiscent of the lilting whistles of the Clangers.

After a few more fruitful viewing hours, I walk back towards the farmhouse and spy Dewi-dog sitting to the side of the coast path, his eyes fixed on the slopes below. When he spots me, he stands, his tail helicoptering a welcome – though he's dying to gallop up to greet me, his gaze keeps returning to the cliff-side as if someone down there has instructed him to sit and stay and he's weighing up whether or not to disobey them. Fortunately, when Lisa scrambles up the cliff to the coast path again, she's too elated at having spotted a rare avian migrant, a yellow-browed warbler, to focus on Dewi's near-misdemeanour.

As we walk on together, I notice that Ramsey Sound looks even more agitated than earlier. It's not surprising that the tidal race, and its riotous meeting with the reef known as the Bitches and Whelps to the south of the harbour, should have acquired such a formidable reputation.

'I suppose you know about the turbine?' says Lisa.

I know that an attempt's been made to harness the power of the sea here and that a tidal energy conversion unit, comprising a base frame and a wind-type turbine, has been installed beneath its surface. I haven't, however, followed the progress of the project.

'There is no progress,' she tells me. 'It's still in there but it's not working.' She discloses that the unit generated electricity for just three months before a fault developed with a sonar, the purpose of which was to detect the presence of cetaceans and monitor their movements around the turbine. Though Ramsey Sound was chosen as the most propitious site from over twenty possible UK locations, the system was no longer permitted, as a result of this fault, to operate within its licence. Later, a mechanical defect, which would have prevented it from generating, was also identified.[1]

I've encountered various kinds of anthropogenic disturbance of seals on my journey so far but this latest example feels like the most contentious. We urgently need, of course, to source forms of renewable energy and the fact that tidal, unlike wind and solar, is predictable is a big advantage. Yet, as the venture's attempt to monitor cetaceans' behaviour recognises, the implications for seals and other marine mammals, including the harbour porpoises here in Ramsey Sound, who favour tide races as feeding grounds, need to be fully evaluated. Collisions with turbine blades will surely occur and the essential curiosity of vulnerable weaned pups, who leave the breeding beaches with no knowledge of prey location or the wider ocean environment, puts them at particular risk. If there are going to be wildlife casualties, how clean and green can we believe tidal energy to be?

At this point in time, I doubt that it's possible to give a definitive answer to this question. Compared with the wind industry, which has seen the installation of thousands of turbines and accrued detailed information about their effects on wildlife over many years, the tidal industry is at an early stage of gathering and interpreting data. While further projects are being trialled at test sites around the UK and a commercial tidal energy scheme is already operational in the Pentland Firth between the Scottish mainland and Orkney, research into their environmental impacts is ongoing. In Northern Ireland's Strangford Lough, from 2008 onwards, a now-decommissioned turbine was deliberately switched off whenever its active sonar detected the approach of a marine mammal and, though this mercifully mitigated collision risk, no information on the extent to which a seal might be able or willing to evade rotating blades could consequently be collected. Thus far, no other published study has provided evidence of close-proximity avoidance of moving blades either, though analysis of Strangford Lough data revealed incidents of more distant evasion, with some tagged seals displaying reluctance to swim within two hundred metres of the operating turbine. I also recall coming across a report, produced by the Sea Mammal Research Unit at the University of St Andrews, that described the carrying out of a preliminary trial – a series of collisions, at a range of different speeds, between an approximation of a tidal turbine blade and five adult and juvenile grey seal carcasses. Post-collision X-rays and autopsies revealed no serious skeletal trauma and the study concluded that less than a third of impacts are likely to end in death. Yet even a minor underwater head injury causing concussion or disorientation could be fatal to a seal who needs to swim back up to the surface to breathe, and the study wasn't able to quantify potential damage of this kind.[2]

So many tidal turbine questions are streaming through the strait of my brain that, when we finally go inside the farmhouse, I can't immediately absorb Lisa's promised explanation about the software she uses for ID-ing seals. She's pulled an extra chair over to her PC in a cosy room with a Welsh flag draped above the fireplace and Dewi cwtched up in front of it.

'The software's called ExtractCompare,' she says. 'You upload your cow photo – ideally, it'll be a wet female lying on her belly so her unique markings are well defined, like this one – then you need to drop points onto the picture.' She demonstrates by clicking on the cow's nose, right eyebrow, right flipper. 'This creates a patch like a fingerprint that it'll compare with others in the database. You leave the computer running for a few hours and then the top ten possible matches come back and you make the final decision about whether it is indeed a cow that's been sighted and photographed before.'

In other words, the software does all the complex and laborious match-finding work that Sue Sayer's trained herself to perform.

'Can you use it for bull seals too?' I ask.

Lisa shakes her head – there needs to be a strong colour contrast in the pelt patterns and bulls, especially as they mature, tend to be more uniformly dark with lighter areas less apparent. 'It doesn't work for scarred animals either. The software can't cope as well as the human eye can with scars that change over time. So I still maintain my own catalogue of scarred animals. Since 2006, I've identified over three hundred, mostly with net constriction scars.'

During the winter months when Ramsey is closed to visitors, Lisa's contracted to input thousands of cow photos taken by wardens at other locations too. 'We get lots of matches so we know that cows are regularly moving between sites. There'll be a seal on the beach here on Ramsey that I think is "mine" but which Skomer is convinced is "theirs". And Sue Sayer thinks a bull is "her" seal, yet it's chosen to come and haul out with us. We really need to stop being like that, claiming seals as "ours". They're all seals of the South West. A big moving population.'

While she fetches a bundle of letters for me to post once I'm back on the mainland later as she won't be heading across the sound for another few weeks, I reflect on what she's just told me. Dropping the possessive pronoun when referring to the itinerant seals of Wales and Cornwall makes scientific sense, yet less than half an hour ago, as we discussed the turbine, she made an emotional exception. When I questioned whether the risk of seals dying from collisions meant the tidal energy unit represented an unacceptable level of human disturbance, Lisa's response was unequivocal.

'I'm not happy to think that even one of *our* pups could have got into trouble with it.'[3]

6

Fleeing Ophelia

Within a few moments of joining the coast path above my home bay of Aber Felin, I spot a dead shrew, its long nose tapering towards a dead slow worm.

It's hard not to think of both as bad omens.

Towards the mouth of the bay, a raft of seaweed and plastic debris is reeling and keeling. A dark shape in the heave of sea in front of the first of the three beaches turns out to be the trunk of a tree rather than a beachmaster bull or a cow overseeing her pup. And judging by the purposeful hurl of the waves, this beach will be taken over by the tide within the next hour, with just a few patches of pebbles at the base of the cliffs remaining.

In my pre-seal-watching days, I used to love hearing and seeing a heavy sea, all the theatrical crashing providing a thrilling resonance with the turbulence I often seemed to feel inside. But now a dramatic sea fills me with dread rather than exhilaration.

Today, especially.

Earlier this week, ex-Hurricane Ophelia stormed northeast from the Azores, rampaged through Ireland, then flung itself at Wales. Before the storm, the wind was weirdly warm, the sun glowed red and the sky yellowed as Saharan dust was blown in. I've experienced plenty of ferocious autumn and winter gales during my years of living in Pembrokeshire but the wind that gained momentum over the course of the afternoon, evening and for much of the night was the most violent I've known. While local attention was focused on downed power lines, blocked roads and fallen trees, I fretted over the potential loss of Aber Felin seals, from the just born to the just weaned, and obsessively checked the Ramsey and Skomer social media feeds for updates on the islands' pups too. Lisa posted photos of monstrous waves engulfing Ramsey's harbour and questioned how the seals on the beaches, many of whom I would have seen when I visited, would ever survive. Bee and Ed on Skomer, sharing photos of the 'boiling' sea, posed the same grim question.

Following these alarming mid-Ophelia updates, more details of the damage wrought by the storm emerged in its immediate aftermath. Two-thirds of Skomer's seal pups were estimated to have died, while many of the survivors, some of whom had washed up on a different beach from the one on which they were born, had suffered multiple injuries. As for Ramsey, in addition to documenting the loss of three-quarters of the island's pups, a blog post drew attention to the plight of the struggling orphans – pups left by mothers who ultimately elected to save themselves by heading out to sea instead of risking death by staying on the wave-thrashed beaches.

This is the first time I've returned to Aber Felin since the storm and I'm feeling seriously jittery about what I may – or may not – find. Last week, it being early October, the pupping season was at its peak, with fifteen pups in moult, several of them on the cusp of being weaned, and a further twelve whitecoats. Even though the bay's north-facing – unlike, for example, Ramsey's exposed, west-facing Aber Mawr – it seems inconceivable that 'my' pups could have escaped the devastation. I know, too, that seal rescues can't be effected here, both because the beaches are so inaccessible and because of the risk of causing disturbance to the rest of the breeding colony, so it won't be possible to alleviate any injured animal's pain.

Whitecoat? No, just a pale-coloured rock.

Flippers of a moulting pup who's mostly hidden by a boulder? No, just some soggy fronds of kelp.

As I scrutinise the first beach, I'm struggling to differentiate what's there from the wishful thinking, to realise that I'm not genuinely smelling, but hallucinating, musky pheromonal bull. I eventually count just five pups, one with a bloody front flipper, and five cows, one with a damaged left eye.

All of these survivor seals seem more edgy than usual and more unpredictable in their behaviour. On one side of the diminishing beach, a pup moves with a baffling sense of urgency towards the sea, prompting its mother to barrel after it and herd it back onto the pebbles. On the far side, it's the opposite story – a cow cannonballing towards the sea, pursued by her pup who struggles over, and half falls off, the larger rocks that block its route, its fur more murky brown than white, suggesting that it's already had some rough moments in the water. It immediately gets buffeted by the waves this time, too, and retreats again, wailing. Its mother comes back onto the beach and lies on her side so her pup can feed, though she opts for a spot that's much too close to the water's edge and waves continue to wash over them. I wonder if she's an inexperienced first-time mother or whether the trauma of the storm has given rise to her rather erratic behaviour.

I slosh along the path, ankle-deep in mud and puddles in places, towards the viewing point over the middle beach, the triangle of pebbles sandwiched between two bluffs. This is never as popular for pupping, but I still recorded a couple of young ones here last week, who should have been approaching peak plumpness by now. There's no sign of them though – where they lay, a misshapen blue crate and a clump of fishing net now sprawl instead.

Rainwater run-off from the fields that flank the path is coursing down the cliffs at the back of beach number three and I watch a whitecoat exploring the portion of this water that's splashing and pooling in the hollow of a rock. Just a metre or so away from this refreshingly normal, inquisitive activity, a great black-backed gull is tearing the flesh from a dead pup. The ragged red of its severed neck is eerily echoed by the frayed red edge of the waxcap fungus in the grass next to my right boot.

Even more distressing is the sight of a scraggy pup with baggy folds of skin that it hasn't yet expanded into, dragging itself across the pebbles, its head disproportionately large, its eyes bulging. It's small as a newborn but much more mobile, desperately so. Crying like a human baby, it's trying to suckle every object and creature, from rocks to other seals, with which it comes into contact. When it tries to mouth a chubby pup's belly, the latter's mother grabs the hungry one's back flippers between her teeth, shakes it and tosses it away. It's agonising to witness this – an agitated orphan craving nourishment, destined to starve.

From my usual ledge lookout, I force myself to look beyond the orphan's suffering, to tot up the number of seals on the beach and enter the total into my notebook. Though there are six pups in all, significantly down on last week's numbers, none are weaners. I'd expect those who were close to being weaned last week to still be around at this stage, but Ophelia perhaps prompted them to move off the beach and take their chance at sea. It feels impossible to believe that, at only three weeks old, they could be strong enough to survive but at least they may have avoided being smashed against Aber Felin's cliffs by the waves. And satellite tracking has, of course, oftentimes shown that just-weaned pups can successfully undertake long, exploratory journeys.

Again. That uncannily human, harrowing cry. The orphan's approaching another pup now – this one curled like a comma, almost fully moulted – and starts mouthing every part of its body, from neck to back to tail. Though the moulter wriggles and flaps its front flipper, it doesn't threaten to bite, and its mother's thankfully more chilled than most cows too, willing to tolerate the hungry one's presence. When she invites her pup to suck, the orphan continues trying to feed from its rear flipper, and there they lie in a chain of suckling.

Peaceful mother.

Contented, well-fed moulter.

Starving pup, fated to end as food for great black-backed gulls.

After my post-Ophelia visit to Aber Felin, I temporarily try to set aside the individual tragedies – the terror of wave-swamped pups, the cows hunting for their young, the battered corpses returned to Pembrokeshire's shores in grotesque parodies of healthy haul-outs – and focus instead on the broader population picture. A reporter on a TV news bulletin blithely claims that in the three years preceding the storm, the three highest-ever numbers of births on Skomer were recorded so the West Wales population can easily tolerate 'one bad year'. But thanks to anthropogenic climate change, the severity and frequency of storms are on the increase. Is the population robust enough to withstand two bad years? Or three? Or more?

The impact of extreme weather events has to be seen in the context of all the other threats that grey seals are facing too. Climate change brings further trials including variations in the availability and distribution of their prey – around large parts of the UK, for example, sandeels, which are known to be sensitive to sea temperature rise as the cold-water plankton on which they feed shift north, feature prominently in the grey seal's diet. In addition to climate change, they're dealing with the cumulative burden of marine debris, toxic pollution and probably a slew of other perils to which I haven't been alerted on my seal journey yet. Since it's impossible to quantify the synergy of all these hazards, I question how feasible it is to believe that after one or more 'bad years', population levels will remain sustainable.

Increased storm severity is also having a marked impact on those who work in seal rescue and rehab as they struggle to accommodate the escalating number of animals who end up in their care. The nearest rescue centre for injured Pembrokeshire pups is RSPCA West Hatch on the outskirts of Taunton, at least three stressful transportation hours away. I first met the centre manager, Bel Deering, when we sat next to each other at an animal studies conference a few years ago and exchanged whispered opinions on the speakers. When I contact her again to wangle a tour of the West Hatch seal facilities, our conversation is conducted at a rather higher decibel level as we strive to make ourselves heard over a pup who's hollering in one of the outdoor pools.

'Some of these pools are designed for oiled seabirds so they have vertical sides,' Bel tells me. 'Now that they're seal pools, it would be more appropriate if the slope was gentler but we're having to juggle different spaces. At one time, we'd get lots of oiled birds but now it's lots of seals instead.'

'How many is lots?'

'Ten years ago, we'd get five or six seal pups over the winter. This past year we had a hundred and fifteen. The problem is that along with the changing weather patterns, many small, private sanctuaries all around the country are shutting down.'

I'm aware that this is the case in Pembrokeshire. Milford Haven Seal Hospital closed in 2014 and though its ex-owner, the tireless Terry Leadbetter, still rescues, and is able to temporarily medicate, seals on his premises while he waits for space to become available at West Hatch, the county is in ever more desperate need of another sanctuary of its own.

We pause at a tall mesh fence which encloses a small group of moulted pups, one of whom lollops in our direction, no doubt associating humans with the arrival of food. 'This is our large waterfowl paddock,' says Bel. 'It's designed for swans but as you can see, it's a large seal paddock at the moment. We've cordoned off the grass 'cause they roll around in the mud and get terribly filthy. There's a deep, good-sized pool in there, though – you can tell they enjoy it.'

They seem to be enjoying, or at least not minding, the tenacious drizzle that's falling too. Though Bel and I both have hoods on our waterproofs, we neglect to use them, as if they'll somehow obstruct the flux of our chat. Already, Bel's long, turquoise hair has a sheen of raindrops while I'll need to resort to drying mine later with paper towels in the loo.

With West Hatch having to make creative use of the limited space that's available, I ask what happens when it reaches maximum pup capacity. Newly rescued seals have to be driven two hundred miles further east to its sister centre, RSPCA Mallydams Wood near Hastings, Bel reveals, or to Stapeley Grange in Cheshire. If both of these are also full, the final option is RSPCA East Winch in Norfolk. Then, at the end of the recovery period, all the Pembrokeshire pups from the genetically distinct subpopulation of grey seals of Wales and the South West who've rehabbed on the east side of the country have to make the journey in reverse so they can be released close to where they were rescued. It's perhaps not surprising that some argue it might be kinder if, when they're called out to undertake rescues, RSPCA inspectors were to eutha-nise pups who are under a certain age and weight in order to spare them both the arduous road trips and three-to-four conceiv-ably fraught months in the alien rehab environment.

The paddock pup continues his hopeful patrol of the fence while Bel, mindful of the importance of not habituating him to humans, battles to avoid eye contact. 'It's really tricky – they're in a confined space and we've got to work with and around them so can't avoid it completely. We try not to talk to them but it's very hard as they're so gorgeous. And their eyes are so hypnotic, like they have some kind of magic powers.'

Tempted though I am to focus, for what's left of my visit, on the notion of seal magic, Bel swiftly switches back to the pragmatic.

'Our water bill's five grand a month at the moment,' she confesses, gesturing towards a cleaned pool that's being refilled, much to the delight of its seal occupants. 'And at peak times, we get through eight hundred kilos of fish in just five days.'

Her words highlight another critical consequence of climate change and I realise that the Cornish Seal Sanctuary's ownership by Merlin Entertainments and provision of visitor attractions that help fund their costly rehab work is far more expedient than I initially understood. As rescue centres endeavour to care for an ever-increasing number of injured and orphaned pups, it's not just lack of space that will be a concern going forwards, but funding too.

'So, Susan. Let's have a little chat about your dad.' The nurse sitting opposite me has a pen in her hand and a form attached to a clipboard resting on her knees. 'Have you noticed any changes in him at all?'

The room is windowless, nondescript. Square tables pushed against the wall. Orange plastic chairs. 'What sort of changes?'

'Well, how's he managing at home at the moment? And when he's out and about – how's he managing with shopping?'

'He still does all his own shopping. He's managing fine.'

She raises her pencilled-in eyebrows, arched like coastal erosion. 'No problems when it comes to paying for things? He can count out his change?'

'Yes, no problems there.' Just that time, in a café, when he held a 5p coin between his thumb and forefinger and stared at it for a while like he'd never seen it before in his life. 'As I said, he's doing fine.'

A box is ticked. A note scribbled. 'And what about his fridge, Susan? Have you ever found out-of-date food that he's forgotten about?'

'Don't think so.'

She frowns. The sea-arch brows collapse.

'Well, maybe the once. A yoghurt. And a Scotch egg. Oh, and there was some cheese that went mouldy. And once I found some broccoli that had gone yellow. Nothing major, though – we all miss use-by dates from time to time.'

Another box ticked or crossed. More words jotted.

At around the time of my Skomer visit, soon after Dad failed to remember having read *Waterfalls of Stars*, he had an annual check-up with his GP who, unprompted, decided to perform a short memory test. Nothing to worry about, Dad insisted, just the normal ageing process, but still, he was referred to this memory clinic to check things out further. Now, while he has a more thorough cognitive assessment in the room next door, the nurse continues to quiz me on how he's coping with all his daily doings. While part of my brain's engaging with our conversation, another part's speculating on the questions he's getting and the answers he might be needing to find. Today's date? The names of the Prime Minister and the President of the USA? The months of the year backwards? A name and address to memorise and repeat ten minutes later? If he's asked to do the latter, I hope his short-term recall will be more effective than that of the seven captive pinnipeds, including a grey seal, who featured in a Danish study. It was concluded that they could remember what they'd just been instructed to do, such as wave a flipper in the air, and repeat it on command, but only if it was within eighteen seconds. Though I'm a little sceptical of these results – the study took place under artificial conditions and the animals would surely fare better in their natural environment, undertaking tasks related to their survival rather than contrived actions designed by humans – I still find myself willing Dad to far outstrip their alleged eighteen-second short-term memory limit.

After another twenty minutes of discussion with the nurse, I head out into the corridor to wait for Dad to appear.

'Susan, before you go, let me give you this. You might want to have a little read.'

I glance at the cover of the leaflet she's handing me. As Dad emerges, smiling, I bury it in my bag.

'Well, that seemed to go okay – I quite enjoyed it.' He pulls his fleece on over his head, repositions his glasses. 'Couple of the questions had me stumped but that's how it goes some days.'

That wasn't one of your newspaper quizzes or crosswords, Dad. It's not a game. I've just been given a leaflet on Living with Dementia.

'All right, Geoffrey?' The collar of his fleece is askew and the nurse untucks it, smooths it down.

He can do it himself. He's not incapable.

'There you go, Geoffrey. That's better.'

There's talk of other tests, a brain scan. An appointment with a specialist when all the results are through. I'm feeling like I'm reeling, like I've got a personal version of Storm Ophelia sweeping through my day.

Strategies for adaptation – that's what I need to focus on. Grey seals to climate change. Bel and her RSPCA staff to the shortage of space.

And me to Dad's likely deterioration.

Issues of resilience and adaptability remain very much at the forefront of my mind in the weeks ahead. Shortly after Ophelia, a second major storm ravages West Wales – though its name, Brian, couldn't be more benign, it savages yet more pups both on the islands and all along the coast.

It's late October, on a showery but calm day, before I have the chance to return to Aber Felin. Each year at around this time, I expect to start to see a shift in the role and occupants of the three beaches. The first two tend to be quieter, populated only by the final few pups and their mothers, while the third becomes busier, gradually transitioning from an autumn pupping spot into a cows' winter haul-out and moulting zone. This year, though, in the wake of the two storms, I'm not sure what I'll find – I'm just hoping to be reassured that there's still a recognisable rhythm to the Pembrokeshire pinniped year.

Cows of the agricultural kind are grazing one of the fields to my right when I follow the path around the bay. They pause in their chewing as I pass, plod towards the barbed wire fence between us, curious but cautious, ready to back off if I make a sudden move. More and more start to cluster at the edge of the field, one pushing her head and yellow-tagged ears between the bodies of two others to watch me. Not surprisingly, meeting her gaze, with its glaze of subjugation, feels so different from being eye to wild eye with a seal.

Once I reach the viewpoint above the first beach, my attention is immediately grabbed by a fat whitecoat sleeping on its back, flippers sticking out at right angles to its body, like a child wrapped up in too much winter clothing. At the back of the beach, an even chunkier pup, deep in moult, is tucked into a fissure in the cliff. Overlaying the usual smells of salt and kelp is the pungent odour of lustful bull and a step closer to the edge brings an attempted aquatic mating into view. The bull's mouth is agape and though the cow, probably the mother of the moulter, slaps the water once with her front flipper, she otherwise seems happy to swim along-side him, gargling and blowing bubbles. At last, he briefly clasps her, his belly to her back, his flippers enclosing her body, but when he tries to sink his teeth into her neck, she twists her head to snap at him, moans a warning and breaks away. For all the hours I've spent here over the past few years, I've yet to witness a successful mating.

'What's happened round here then – why have you lost so many seals?'

I'm joined by a woman in a Help for Heroes hoodie who tells me she's on holiday from Kent and whose query about the seals implies she thinks I've carelessly misplaced them. As I fill her in on the story of the storms, she listens with her head on one side. 'Poor buggers,' she says when I've run out of words. 'I read it was on course to be a bumper year for pups. So the storms must be nature's way of controlling the numbers.'

This argument – describing something as 'nature's way' when science suggests that human-induced climate change is a major driver – thoroughly irks me and it's not the first time I've heard it used over the past month. Yet for all my irritation, I can under-stand why it's so appealing. If a tragedy can be explained away as 'natural', it absolves us of responsibility, palliates the sadness, makes it so much easier to bear.

While the woman heads west, I move on to beach number three. As well as a suckling, nearly weaned pup, I count eleven cows, early recruits to the winter haul-out, who may have already raised – or lost – their pups elsewhere around the Pembrokeshire coast. For the next half hour, apart from these seals, a pair of juveniles rolling in the shallows and the bull from the first beach now bottling just offshore, my only companions are a heckling herring gull and half a moon. But then a bespectacled walker, his balding scalp slick with sweat and sprinkles of rain, shuffles up. When he extricates himself from his rucksack and dumps it at his feet with an outbreath of relief, it's clear he means to keep me company for some time.

He's visiting from Trim in Ireland, he tells me. 'I'm in a walking group for middle-aged men and we do six kilometres a week.' Though it's deliciously apt that he's seeking to improve his fitness in a town called Trim, I wonder how well his weekly walking has prepared him for the demands of the Pembrokeshire Coast Path. 'It hasn't,' he acknowledges glumly, 'I haven't got the miles in me legs.'

When he moves on to talk about his dairy farming day job, he's visibly cheerier – 'I milk a herd of one hundred and ten and that's just enough for a man of my age' – and thereafter, every time I share a snippet of information about the breeding behaviour of seal cows, he draws a comparison with the behaviour of cows of the bovine kind. He has an endless succession of questions too, addressing me as if I'm the seals' keeper and they inhabit a curious midway space between domestic and wild. 'How long are the females fertile for? With my cows it's sometimes only fifteen minutes and I've got a machine that tells me the best time for insemination. D'you have any problems with placentas that don't come away or stillbirths or other things that need vet intervention?'

While we've been chatting, the bull's made his way onto the beach, bypassed the group of howling and growling cows, and ploughed on towards a seal with leopard-style rosettes spotting her flanks. She's the mother of the nearly weaned pup and therefore likely to be the only seal on the beach who's in season. She seems very relaxed at his approach, yawning rather than flipper-flapping.

Wiping drizzle from his glasses with the sleeve of his fleece, the Trim Man turns his back on the beach and looks inland at the fields instead. The sight of a tractor spraying lime inspires rather more excitement than the prospect of watching a potential mating.

The bull seal snorts, manoeuvres himself alongside the cow, slings his front flipper across her back and presses his body against her. They rest in this position for a few moments, then she wriggles out of his grasp and moves towards the water. His thick, pink penis flops against the pebbles, then retracts, as he follows her into the shallows, where they twist and twine in a more purposeful, adult version of the ongoing dance of the juveniles. At last, he moves behind her again, clasps her with his flippers and buries his teeth in her neck, and there they stay on the wave-washed edge of the beach, her back to his belly, in stillness. I'm finally seeing a proper copulation for the first time.

'Blimey,' says the Trim Man, having turned towards the sea again, 'he's taking his time.'

'Mating's been known to last for up to an hour,' I say.

'An hour!' he echoes, whether in admiration or disbelief it's hard to tell. 'An hour!' He continues to express his consternation as he heaves his rucksack onto his back, nods a farewell and resumes his shamble along the path.

As for me, I stay observing the beach beyond the moment when the cow and bull separate and beyond the end of the misty rain. Though the sky's still mostly overcast, there's a warmth in the air that I'll long for in the months ahead when the conglomeration of cows reaches its winter peak and my fingers are too numb to hold a pen.

'Lovely day!' A trail runner, shins splattered with mud, and fizzing with endorphins, suddenly whizzes past me.

'Fabulous!' I say, though we both know that, meteorologically speaking, it's nothing of the kind.

But from a seal-watching point of view, it's resoundingly so. While a single triumphant mating can't compensate for the loss of hundreds of this year's pups, it's proof that Pembrokeshire seal life is enduring. And it inspires a buoyancy I hope I can draw on as I accompany Dad through his cognitive decline.

7

The Red Wilderness

Clamber uphill. Edge my way down. Repeat.

Under a mix of sun and whitecoat clouds, a welcome spell of late October weather, I'm undulating along the most remote section of the Pembrokeshire coast, no more than twenty miles east of Aber Felin. In addition to the present-day thrill of the formidable cliffs overhanging inaccessible shingle beaches, the area is historically significant, as it's where naturalist Ronald Lockley pursued his trailblazing grey seal research in the 1940s.

Y Godir Goch, he called it. The Red Wilderness.

Born in Cardiff in 1903, Lockley moved to Pembrokeshire at the age of twenty-four to take a long-term lease on Skokholm, the island I spied from Skomer. He began researching migratory seabirds, including the Manx shearwater, and started to publish articles and books in a range of genres from scientific monographs to autobiographical accounts of island life. He also succeeded in luring filmmaker Alexander Korda to Pembrokeshire to produce an early natural history documentary, *The Private Life of the Gannets*. Scripted by Lockley, it was awarded an Oscar in the Best Short Subject, One-Reel category.

Forced to leave Skokholm after the outbreak of the Second World War when it was requisitioned by the military, he settled and farmed on the Pembrokeshire mainland. One of the books he

1

subsequently wrote, *The Private Life of the Rabbit*, later provided his friend, Richard Adams, with insight and inspiration while working on *Watership Down*. Writing aside, in addition to playing a principal role in the setting up of the Pembrokeshire Coast National Park, Lockley was instrumental in the mapping of the county's coast path, now a National Trail, which I access and appreciate almost every day.

Over the course of his long life, he published over fifty books, the last of which, a collection of letters, appeared in his ninety-second year. His science is rigorous yet imaginative, born of living alongside his animal subjects, qualities that are especially prominent in *The Seals and the Curragh*, which details his Red Wilderness research. After a perilous solo sail to the stretch of coast along which I'm now walking, Lockley set up camp on a pupping beach for several weeks, sleeping beneath his upturned boat by night and carrying out his ethological studies by day. He's justifiably vague about the exact location of his camp 'for among the fishermen and week-end sportsmen there was at that time a local organization for the purpose of destroying the seals', but by orientating myself via an Ordnance Survey map and Lockley's topographical descriptions and photos, I hope I'll be able to figure it out.

A straggly, V-shaped skein of Canada geese flies over, following the line of cliffs, leaving me plodding in their wing-whumping wake. A raven clocks my presence from a fence post and a stone-chat chak-chaks his alarm call from a twist of barbed wire. In spite of the fact that occasional Private/*Preifat* signs are keeping walkers without, and livestock within, the network of fields to my left, Lockley's portrayal of this area as far removed and un-frequented still feels true today. His references to the land mammals that can be found here are equally tantalising – not just badgers, foxes and otters but also polecats, a species I haven't yet managed to spot.

While his broad landscape descriptions – the 'high, north-facing cliffs', steeply shelving pebble beaches and 'good, rocky feeding grounds' – chime exactly with what I'm currently experiencing, I need more specific details if I'm to precisely pinpoint the location of his camp. I pause at a stream that's spurting over the cliffs to check the OS map and see that 'Godir' appears in the names of several nearby kinks and inlets – Godir-y-Golomen, Godir Rhyg and Godir Tudur – which points to my being in the right area. I try to match the rocks and crags in Lockley's black-and-white photos in *The Seals and the Curragh* to the scene before me too but the fact that my viewpoint's over a hundred metres above his sea-level perspective makes this quite tricky.

If only I had access to a copy of the West Wales Grey Seal Census, I'd be able to confirm if this immediate area is traditionally populated with pups. Commissioned by the erstwhile Countryside Council for Wales, it's a hefty report, published following sea-borne surveys of 225 beach and cave sites in three consecutive breeding seasons in the early 1990s. When I visited Ramsey, I had a brief glance at Lisa's much-thumbed copy but didn't have time to work my way through its several hundred pages of graphs, tables, maps and analysis, and I've subsequently had no success in tracking down and securing a copy of my own.

My piecing together of limited clues from the OS map and *The Seals and the Curragh* is therefore as close to a categorical answer as to the whereabouts of Lockley's camp that I'm going to get. I can't see the full extent of the beach below because of the angle of overhang of the cliff, but still scooch to the very brink. Though I spy just one weaner, nosing clumps of spume like cottage cheese at the shoreline, I can hear some chuffing and pebble-shuffling, suggesting there are several other seals beneath the overhang, out of sight.

Lockley began to familiarise himself with the gathering of seals that was birthing, suckling and mating here as soon as he arrived. As was his practice with the animals he studied, he gave the seals names, either inspired by their fur patterns like Sue Sayer – a cow with chest spots resembling a rosary became known as Nun – or by a particular incident or behaviour, so that a pup born right next to his boat was called Curragh.

Often employing hands-on research methods that would be frowned on today, he meticulously monitored each stage of Curragh's development, weighing her on livestock scales hung from a stand formed from the oars of his boat. Though 'she struggled wildly at first', he spent hours talking to and 'taming her until she no longer snapped at me, but at last dozed away under my relentless stroking of her white pelt'.

Interesting though it is to follow Curragh's progress, the emotional heart of the book lies in the narrative of a seal who was only two days old when Lockley first accessed the beach and whom he named Billy. Realising that the pup had been abandoned and would die without intervention, he somehow managed to climb from camp to cliff top and walked to the nearest farm. There, he met thirteen-year-old Tessa and arranged for her to bring half a gallon of milk to a drop-off point above the beach every day. Later, he agreed to let her join him on the beach and she even brought along a goat from whom Billy willingly suckled. No less surprising than this goat–seal surrogacy is Tessa's ability to negotiate a route down the bracken-clad cliff. From where I'm precariously perched, it looks almost impassable.

With several more features of the Lockley landscape to locate before dusk muscles in, I retreat from the edge, stretch stiffened legs and continue walking the Red Wilderness. En route, I glimpse a couple more seals – a moulted pup swimming on his back, a cow gliding to the shore through wafts of wave and kelp. It's hard to reconcile today's sedate sea with Lockley's dramatic account of sailing his wind-lashed curragh along here when bound for the seal beach for the first time. I note that versions of some of the landmarks to which he refers – 'the sentinel Careg Rock' and 'the Trwyn', the headland he rounded – exist on the OS map, though his implication that the Red Wilderness lies some hours from the Trwyn doesn't reflect the actuality. The challenging sailing conditions must have made it feel like hours and it may also be deliberate obfuscation to preserve the secrecy of the location.

Obfuscation and secrecy also played a role in a drama that unfolded in the coves of this coast in the heatwave summer of 1983. Keen-eyed local fishermen alerted the police to the suspect presence of several strangers on two Red Wilderness beaches who variously claimed they were testing equipment for an expedition to Greenland and making a film about seals. On one of the beaches, a man-made, watertight, underground chamber was discovered which ultimately led to the exposure of an international drug smuggling operation. Referred to as Operation Seal Bay, the investigation took the police across the Atlantic, as well as to the Channel Islands, the Isle of Man and the south of France. Eight felons were arrested including a millionaire Danish narco-gangster who'd been pursued trans-globally for the past eleven years.

At the joint trial of three of the men, the judge praised the inhabitants of the nearby settlement of Newport and its surrounds for their vigilant response to instances of suspicious behaviour, and it's Newport's dune-fringed beach that I drop down to now. I'm hoping to spot some of the Portuguese men o' war that the tide has recently been delivering to beaches all around the Pembrokeshire coast, but I walk the full length of the strand without seeing any.

I can, however, see the houses of the historic Parrog harbour to the west of the beach, which is possibly 'the little port of X' to which Lockley considered escaping at the time of a looming equinoctial storm. Curragh, by now successfully weaned, headed out to sea when the storm arrived, though at least two younger pups remained on the Red Wilderness shingle along with their frantic mothers. Still worse than reading of their distress is learning what became of Billy. Though Lockley took the decision to carry him up onto a heathery mini-headland above his camp, securing him as best he could, he either bit through or broke his tether and tumbled over the edge into the seething water. Post-storm, having sailed to 'a lee shore where the current and savage north wind' would be expected to drive the afflicted seals, Lockley found Billy's 'meagre body'.

Where the Seals Sing

As my feet turn towards the cafés of Newport and my belly towards the prospect of a hunk of cake, I ruminate on the remaining sections of *The Seals and the Curragh*. Compared with the emotional intensity of the Red Wilderness pupping season chapters, what follows is more restrained in tone, yet it remains very informative. It includes details about the rhythm of the seals' year in West Wales and comparisons with colonies in the rest of the UK, as well as an overview of global grey seal populations and an analysis of the threats to their survival. And even though a few of the facts that Lockley presents may now, decades later, be challenged and re-interpreted in the light of advances in science and ethology, the overall significance of his contribution to grey seal studies is considerable.

An unexpected consequence of my quest to discover the location of Lockley's camp is a rather conflicted feeling in the days that follow. While I of course believe that it's ethically essential to watch seals from a distance and cause no disturbance, part of me still feels frustrated at having been confined to a cliff top when Lockley was not only based at beach level but gaining the trust of, and interacting with, the seals alongside whom he was living. But since it would be deeply improper to seek to establish a meaningful connection with a wild seal myself, I try to satisfy my curiosity about the possibilities and limits of the human–seal relationship by binge-reading several books that foreground this theme instead. All were published between the late 1950s and mid-60s and they vary greatly in terms of tone, insight and agenda.

A Seal Flies By, R. H. Pearson's account of bringing two common seal pups to live in his ornamental lake in Wiltshire, offers little in the way of vicarious pleasure. 'The idea first hit me at breakfast,' he flippantly informs the reader. 'I had suddenly realised that I must have a seal.' In order to fulfil this whim, he accompanies a London Medical College professor on a research trip to The Wash. A group of adult common seals is slaughtered on a sandbank to furnish the professor's latest research project and Pearson 'saves' two pups whose mothers are among those shot.

At this point, the magnitude of what he's setting out to do seems to dawn on him – 'There I was, *in loco parentis* to two flippered children of the deep, and just beginning to realise how little I knew about bringing them up.' Even planning in advance what to feed the pups, whom he names Flipper and Diana, seems to have been beyond him. Having brought them home, it occurs to him that there are no high-fat foodstuffs like margarine and cod liver oil in the house so 'They would have to do without it tonight.'

When Diana proves to be more feisty and bitey than he bargains for, he packs her off to none other than Ronald Lockley at his Pembrokeshire estate. After the more placid Flipper's untimely death, though, he welcomes Diana back in typically self-indulgent fashion – 'The children liked having a seal about the place. The lake seemed unnaturally still without one.'

A series of absurd experiments ensues. Having been trained to wear a collar and lead, Diana is harnessed to a 'little white boat', which she tows, with Pearson standing in it, across the lake. However, when he transports her to Devon in the hope of capturing footage of her swimming in the sea, she escapes from her halter and sensibly refuses to be caught again. The book ends with Diana at large somewhere off Dartmouth and Pearson hoping his narrative might inspire others 'to train young seals washed up during storms'. It's hard to imagine, though, how it could inspire anything but discomfort and dismay.

I justifiably approach *Atlanta, My Seal* by H. G. Hurrell with some trepidation. It features another young female seal, a grey this time, who's kept in an outdoor swimming pool by another middle-aged man, far from the sea on the edge of Dartmoor. Thankfully, H. G. Hurrell is an experienced naturalist who's previously kept otters, pine martens, badgers and polecats and though the practice of domesticating wild animals can't be condoned, he at least has a track record of providing long-term care and compassion. Atlanta is a rescued seal too, not one he captures to satisfy a whim: at just a few months old, she's found, wounded and emaciated, in a local estuary.

From the outset, Hurrell is at pains to emphasise Atlanta's capacity for learning, asserting that none of the other animals he's kept can match her abilities. However, some of the tasks he encourages her to master are verging on the ludicrous and he shows a worrying tendency to judge her success based on human accomplishments rather than to recognise her seal-intelligence.

To begin with, he teaches her commands in much the same way as we train our canine companions and, over time, she develops a 'sizeable vocabulary', responding to words such as 'down', 'clap', 'ball', 'away', 'left', 'right', 'roll', 'no' and 'hoop'. But then he embarks on an altogether more bizarre mission – 'I started to teach Atlanta to read.' A photograph showing Hurrell in a jacket, flat cap and wellies, holding up the word 'Roll' in giant white letters on a black background while Atlanta turns onto her back at the side of her swimming pool provides visual confirmation of the unorthodox path he's opted to follow.

From reading to arithmetic. Hurrell's diary of 1962 outlines some of the frustrations he experiences, having decided that Atlanta needs to learn to count, and it includes such risible comments as 'Atlanta seemed to make no effort to become even a modest mathematician.'

By the end of the book, I'm left with the impression that Hurrell is well-meaning but misguided. He's succeeded in raising a physically healthy and friendly seal, though releasing Atlanta back into the wild as soon as she'd put on weight and her wound had healed, as sanctuaries do today, would have been kinder. He does, however, offer a persuasive explanation in support of his decision not to release her in the unexpected form of R. H. Pearson's Diana. Following the publication of *A Seal Flies By*, it appears that she survived in the waters off South Devon for eighteen months after her escape, only to be deliberately 'killed by some wretched fellow out in a boat near Torquay. . . Diana, having grown to trust humans, would have been unprepared for the attack which cost her life.'

It's with some relief that I turn to *The Seal Summer* by Nina Warner Hooke, which, in both content and writing style, immediately sets itself apart from the other two books I've read. Warner Hooke draws on her professional playwriting skills when crafting this prose account and, refreshingly, it's not a story of a pup being raised in captivity. Instead, she describes how, over the course of one summer, a male grey seal about eighteen months old visits a beach in Dorset, voluntarily interacts with the locals, and enriches their lives in ways that none could have imagined. 'From earliest times man has captured and trained animals to serve him,' she rightly says. 'But the instances of a wild animal associating freely and spontaneously with humans are rare indeed.'

In addition to the 'joyous comradeship' that the seal, who becomes known as Sammy, offers, drawing 'together in affection for him people of all ages and types', Warner Hooke humbly acknowledges that she learns a great deal from him. She references Hurrell, describing his relationship with Atlanta as that 'of master and pupil which in my case was reversed'. In fact, when she's asked what might be motivating Sammy to stay in the cove, she replies that he's 'doing some research on the human species'.

Her initial contact with him triggers unexpectedly profound feelings – she experiences it more as a reunion – and the same sense of their encounters being 'a kind of recognition' comes when she first enters the water alongside him. She rejoices in his reaction to her swimming – he seems to move from astonishment that she's not just a land animal to excitement that she's able to join him in his element.

With her playwright's instinct for what constitutes a strong narrative arc, soon after describing the most moving of their bonding episodes, involving hours of play in a sea cave, Warner Hooke introduces a malign outsider. This Canadian, who claims to have been attacked by a seal further west along the coast, wants to seek revenge on the innocent Sammy. *The Seal Summer* builds to a nervy climax as Warner Hooke and her companions work to outmanoeuvre the Canadian and prevent Sammy from being shot. Having finally managed to secure his safety, she draws the reader's attention to more widespread abuse of seals, condemning human violence and profligacy with an urgency that feels decades ahead of its time.

Before Storm Ophelia wrought her havoc, I managed to explore one more seal site on the Wales section of my journey – the settlement of Cwmtydu, an hour's leisurely drive northeast of the Red Wilderness on the coast of Ceredigion. Now that I've read *The Seal Summer*, my thoughts keep returning to my experience of visiting this intimate, accessible, little bay. Snuggled between two headlands and with a river purling into the sea on the southern side, Cwmtydu's beach is the only safe haven for seal pups in the area. On a very high tide, when the nearby pupping cave is flooded, or if the weather's especially turbulent, an occasional cow swims her pup to the beach, and there they tend to stay for the remaining, pre-weaning weeks of feeding.

Cwmtydu also happens to be a haven for humans. At the sea wall between beach and car park, I met Pauline Bett, who, every year throughout the pupping season, organises a rota of volunteers to try to persuade the walkers, swimmers, kayakers, anglers and casual visitors who seek to access the beach to keep a mindful distance. This year's principal pupstar, already close to wholly moulted, was sleeping soundly on a seaweed mound only twenty metres away from us while his mother swam just offshore.

I sat on the sea wall next to a donation box for seal badges, and postcards weighted down by pebbles. About thirty other people, including several photographers with tripods, had gathered to watch here too, thanks to the local newspaper having run a 'Life's A Beach As New Seal Pup's Spotted!' story.

Green-fleeced and freckled, Pauline explained that she divides her time between her paid work as a first-aid trainer and her unpaid monitoring of the beach and its visitors. Her involvement began in 2001 when she came to Cwmtydu to kayak and found a crowd of fellow kayakers on the beach alongside a pup, with just one woman trying to encourage them to be more considerate. A particularly bolshie man ignored her and entered the sea, barging between the pup and its mother. Disturbed by what was happening, Pauline opted not to go kayaking after all, prompting the woman to burst into tears of gratitude. They waited together until the male kayaker returned, at which point Pauline told him exactly what she thought of his insensitive behaviour. 'And from that chance encounter, I got involved and started organising.'

Cwmtydu Bay Wildlife came into being that same year. The first few seasons, in spite of some supportive gestures from Ceredigion County Council, were often rancorous and divisive with a number of locals objecting to the fact that access to the beach was being limited. 'A café banned me for a year!' said Pauline, shaking her head as if still unable to believe it. 'And the owner of the B&B claimed we were spoiling her business, stopping her visitors from getting close to the seals. Now we have a good relationship with the B&B – it's run by different people who encourage visitors to watch more respectfully.'

'Hello, Gorgeous!' A wetsuited woman with a gush of blonde hair hopped onto the sea wall. Thirty hopeful heads turned in her direction. It transpired, though, that she was addressing the bull seal who'd just appeared, plunging in and out of the surf and lunging at the pup's mother.

Pauline told me he's well known and loved by regular visitors to the bay, having been Cwmtydu's beachmaster for a good number of years. 'He came to mate with our cow twice yesterday – it went on for a few hours.'

'A few hours!' Ribald laughter broke like a wave from a pair of elderly ladies with wind-resistant perms. 'Imagine that!'

A cry cut across their chuckles and chat. The pup had woken up, his hunger call deeper and hoarser than that of one recently born, his whole body contracting, rear flippers lifting with the urge to make himself heard. His mother was still swimming, though, so he shunted himself down the beach towards her and let out another yawp. As soon as she heard him, she clattered out onto the pebbles to enable him to feed. At this near-to-weaning stage, feeding might only take place every seven hours, as opposed to the six-times-a-day pattern that's needed by a newborn, though this cow was deemed to be rather more conscientious.

'She's such a good mum, this one.'

'Ever so attentive.'

'Better than some we've seen.'

'Oh, some have been awful flighty.'

The entire time the pup was suckling, many of those watching, the tightly permed ladies included, passed judgement on the cow's suitability as a 'mum' in much the same way as the tabloids do with humans.

'Soon as he's awake, she's there.'

'There in a wink.'

'Never mind the bull and his demands.'

Over the next few hours, once the pup had sunk into sleep again and his audience into silence, Pauline had several potential disturbance incidents to deal with. First, two anglers, laden with bait boxes and rods, marched towards the beach. 'We know how to fish around seals,' they insisted as Pauline's co-volunteer tried to intercept them. Only when a child in a unicorn headband yelled, 'But you're not ALLOWED to!' were they unexpectedly reeled in.

Later, a young couple tried to sneak onto the beach to get a close-up selfie with the pup. When challenged by Pauline, they argued they didn't see the notices warning them to keep away. Considering that there were two huge info boards on the pebbles, one of which was bilingual – '*Morloi Bach*/Seal Pupping in Progress, *Rhybudd*/Caution', framing an exclamation mark in a yellow triangle – this felt like a spurious claim. It would have been hard to miss, too, the permanent Ceredigion County Council sign at the edge of the beach, featuring a cartoon pup with speech bubbles – 'Do not chase me into the water – I can't swim very well.'

In between incidents, I heard about some of the other pressures that the Cwmtydu seals are facing. Every day at this time of year, umpteen tour boats from nearby New Quay cruise along the coast. In accordance with the hierarchy of marine mammal-watching, most of these trips focus on spotting bottlenose dolphins and harbour porpoises, but if the cetaceans prove elusive, attention turns to the third of Cardigan Bay's Big Three, the grey seal. While boat skippers are mostly considerate, I heard that, occasionally, they sail too close to mothers and pups. The appearance of a local hunt at Cwmtydu also caused annual controversy. With their penchant for riding onto the beach and into the sea, participants' behaviour towards both seals and volunteers had, at times, been deeply offensive, as had their denial of all disruption in spite of photographic evidence.

As I continued to listen, the pup woke, yawned, did some front-flipper excavating of seaweed, then fell asleep again.

'What's he doing?'

'Chilling.'

'Lazy little bugger.'

This comment, from one of the tripod-toting photographers, was laced with affection, as was all the proprietorial concern I heard Pauline and co. express towards 'their' seals during the course of the morning.

Thinking back on it almost a month later, I realise that, each pupping season, Cwmtydu offers a diluted, look-but-don't-touch version of Warner Hooke's experience with Sammy, surely the nearest and most ethical approximation to a relationship with a wild seal it's now possible to get.

8

The Dawn Song of Middle Eye

Checked the tide times. Memorised the two-mile route. Won't get sucked into trying to trek to the islands via a shorter crossing – it's mired with quicksand and gullies. Checked the tide times again.

In scant dawn light, I'm tentatively squelching across sand towards the trio of tidal islands, Little Eye, Middle Eye and Hilbre, at the mouth of the Dee Estuary off the Wirral Peninsula near Liverpool. High tide was just before 2 a.m. and I made sure to strike out from the slipway at West Kirby no earlier than five, the start of the recommended safe crossing time. The sand, though, is quaggy and yielding and I can't shake off the feeling that with one misstep, I'll be swallowed up. Unfortunately, there are no footprints to follow – although, in part, it delights me to recognise I'll be the islands' first visitor of the day, it would also be reassuring to know that I'm on the right route. In the absence of footprints, I splosh randomly through hundreds of scattered lugworm casts, the little mounds coiled like the knots – granny, slip, reef, sheepshank – that Dad taught me to tie as a child.

Pause. Quit looking down. Make sure that Little Eye's straight ahead or better yet, just to the right. Don't veer off diagonally towards the other islands.

Having to make such an early start, pre-sunrise, to coincide with the receding tide hasn't been in the least bit inconvenient. I rarely sleep long and deep these days so might as well be exploring the estuary shore as lying in bed, rigid and fretful. It's late spring – over half a year has passed since the West Wales pupping season and in the intervening months, I've been on hand to offer Dad more support as he navigates the sinking sand of his cognitive terrain. Throughout this time, I've followed the usual advice for achieving easeful sleep, from limiting evening screen time to wallowing in warm baths to upping my hours of yoga, all to no avail. I have, however, stopped short of adopting Pliny the Elder's insomnia cure – a seal's right foreflipper has 'a certain soporiferous influence', he wrote in the *Natural History*, 'that, if placed under the head, it induces sleep'.

As I approach Little Eye, an oystercatcher pipes the sun into the sky backed by the skirl of a curlew. I crunch across cockle shells and climb onto the island – it's little more than a grassy knoll with the remains of a brick structure at one end – then pause again to remind myself of the directions for the next stage of the walk. At Little Eye, turn right. Then head across the vast expanse of estuary sand, keeping the next island of Middle Eye ahead and slightly to my left.

The sun climbs higher, looking like a large orange marker buoy. While the extra light and warmth are welcome, the sense of isolation on this stretch is unexpectedly intense. Apart from the islands, there are other landscape features by which I can orientate myself – the coast of North Wales, a troop of wind turbines – yet it's hard not to feel bewildered when walking on such a capricious surface.

It's not a comfortable feeling, yet it's one to which Dad's now having to adjust pretty much all the time.

'Dad! Are you there? Can you hear me? Pick up the phone!'

This is the fifth message I'm leaving and my voice is becoming increasingly shaky and shrill. It's 9 p.m. on a rainy April evening, about a month before my trip to the Dee Estuary. Every night at half-seven, just before he starts to make supper, I give Dad a call to check that he's okay. Some of our conversations have been a bit ill-humoured of late – 'Stop cross-examining me,' he says tetchily, when I ask what book he's reading or what birds he's seen in the local park, 'What d'you need to know that for?' – but he always picks up the phone, always answers.

Today, he had a 5 o'clock appointment with his podiatrist, about twenty minutes by car from his home. Unexpectedly, he's been deemed medically fit and safe to drive and just had his licence renewed for another year, and though I've finally got him to agree never to drive all the way to Pembrokeshire again, he still takes himself on short, local jaunts. I called him in the early afternoon to remind him about his appointment and apart from grouching at me for phoning when he knew all along that it's Feet Day, he sounded happy and well. He should have been home by 6.30 at the absolute latest. Where is he?

There have been quite a few minor incidents in the months since he got his formal diagnosis of early-stage vascular dementia. Lost wallet. Forgotten PIN. A missed dental appointment. Greeting me with surprise when I visit because he claims I never arranged to come. But nothing like this.

My agitated brain hauls itself over the possibilities.

He's in an enviably deep sleep and the phone hasn't woken him.

He's had a fall – he can hear the phone but can't reach it.

He's collapsed from a stroke or heart attack.

He's had a car crash.

He's on life support.

Or worse.

Hooper senses my distress, nudges my leg and I fondle his ears as I stew over what to do. I've tried phoning Dad's mobile but he never has it switched on and probably forgot to take it out with him in the first place. I cast around for someone else to phone but I'm an only child, as is Dad, so neither of us has siblings I can turn to.

What about a neighbour? The elderly man who lives two doors away. Dad's always been so self-sufficient and hates bothering people so doesn't have much neighbourly contact but I know they chat about sport and dogs and the war from time to time.

'Can't get hold of Dad . . . sorry to trouble you . . . been trying to phone him for hours,' I burble when I've managed to track down his number. 'Could you please check . . . I know it's late . . . but could you see if Dad's car's in the drive?'

It'll take him five minutes at least. Coat on. Fumble with the buttons. Stiff-hipped walk out into the cul de sac and down Dad's drive. Ten minutes if the car's there and he rings the doorbell, then walks round the house to try and see what's going on inside.

I open the back door, gulp in air, carry my phone with me into the garden. The rain's stopped at last and I have a weird, befuddled interlude of counting stars as if they're the seals at Aber Felin. Rustle near the compost heap. Fox? Badger? Rat? Call and response of a pair of tawny owls. When my phone finally rings, it makes me jump so much I almost hoot a *to-whoo* myself.

'It's not there, his car. No sign of him. No sign of him at all.'

What now? Head over there? It's a three-hour trip and what would I do when I arrive? Follow every possible home-to-podiatrist route and see what I find?

Should have tried harder to persuade him to move. Sheltered accommodation. Somewhere nearer to here. Should have insisted. He's been adamant that he wants to stay where he is, though, has vivid long-term memories of discomforting childhood moves, wants only what's routine and familiar. His GP agreed that if Dad can remain in his own home, it will be cognitively beneficial too. How to strike a balance between helping him maintain his autonomy and making sure he stays safe? Am I guilty of favouring the former since the longer the semblance of independence is sustained, the longer I can deny the reality of his decline?

Not the time for self-flagellation. Do something. Think.

More scratching of Hooper's ears as I veer between fearing I'm overreacting and knowing I can't afford not to. I reach for my phone again.

The next call I make isn't to Dad's neighbour or answerphone or mobile.

'Police, please. I need to report a missing person.'

A cackle of black-headed gulls accompanies my arrival at Middle Eye. Although, at just three acres, it's small-scale in size, its striated sandstone cliffs are large-scale striking. It is, however, the promised view from Middle Eye, rather than the island itself, that triggered my pilgrimage to the Wirral. Flanked by the waters of the Dee and backed by the North Wales coast is the low-tide span of the West Hoyle sandbank. It's currently peppered with lots of pale dots – more gulls, I guess, though on account of their being so far away, I can't be certain of the species – and bigger silver, cream and black blobs, whose identity is without question.

Atlantic grey seals.

I've never seen such a large gathering in May before. At Aber Felin, most of the seals move on in March following the breeding and moulting seasons and numbers are thereafter low until autumn. Here on the West Hoyle sandbank, though, the opposite is true – it's never supported an autumn breeding colony and the highest counts are recorded in late spring and summer, on rare occasions peaking at over six hundred animals. It seems possible that some of my Aber Felin seals, and others who breed around the Pembrokeshire coast and at smaller sites in North Wales, haul out here, using it as a base to which they return from feeding trips over the late spring and summer months, before migrating back to their autumn pupping locations.

Though there are seasonal and geographic variations in the grey seal's diet, studies over the past three decades have indicated that sandeels, and gadoids such as haddock, whiting and cod, are the most consistently favoured prey, while flatfish like plaice and sole are also popular. Analysis of scat retrieved at known haul-out spots after seals return from their feeding forays enables scientists to reach these conclusions as prey species can be identified from the otoliths – calcium carbonate stones in the ears of bony fish – and the beaks of octopuses and squid, that survive digestion. Mean daily fish consumption per individual seal has been estimated to be between four and seven kilos, though these figures fluctuate according to the size of the seal and the prey's fat content.

There have been sundry studies involving the tagging and tracking of grey seals over the past several decades too, as a result of which some of the secrets of their at-sea movements, feeding activity and physiological processes are starting to be revealed. Whereas they may congregate in large numbers on land, especially at pupping and moulting times, they disperse widely at sea and lead quite solitary lives. In the 1960s, comparatively cumbersome contraptions were used for pinniped tracking that often required the seal to be strapped into a harness, causing hydrodynamic drag with energy expenditure implications. Today's devices are smaller electronic packages in watertight, pressure-resistant cases that are glued to the fur at the base of the animal's neck and stay in place until the time of the annual moult. Tags that beam signals to satellites from which location data is collected by receiving stations are used, as are GPS tags which link into the mobile phone network, yielding such information as length and depth of dives, heart rate, body temperature, swim speed and the distances travelled on feeding trips.

Successive research projects have proved that grey seals display huge variance in their movements. As could be the case here at the West Hoyle sandbank, there's evidence that some make repeated, short-duration journeys of maybe three to ten days to specific feeding areas from a single haul-out site. Others are known to undertake long-distance excursions, swimming for hundreds of miles between different haul-outs, their often straight-line travel indicating an excellent navigational sense. Flat-bottomed, as opposed to V-shaped, dives are interpreted as foraging behaviour at, or near, the seabed, the preferred burrowing habitat of sandeels. Most adults feed at depths of less than a hundred metres and may remain underwater for up to twelve minutes, though they're well able to dive deeper and stay submerged for longer periods of time.

For all these facts and figures, an oceanic gap in knowledge remains where many of the sea-based aspects of seals' lives are concerned, especially with regard to the relationship between their foraging and pupping sites. Because grey seals are capital breeders, fasting and living off their stored resources while they're at the breeding colony, the preceding months' foraging determines both the stamina levels of the bulls and how well cows are able to nourish their newborn pups. Environmental degradation in, or displacement from, their feeding areas could consequently have a detrimental impact on their reproductive success.

As I consider the correlation that could exist between the foraging opportunities beyond the West Hoyle sandbank and breeding outcomes on Pembrokeshire's beaches, I estimate, through binoculars, that today's haul-out comprises about two hundred seals. Due to the distance, I struggle to assess the mix of juveniles, adult cows and bulls, but become aware, while counting, that a chorus of unearthly howls has started to unfurl. In addition to howls, grey seal vocalisations have been described as wails, moans, roars and, extraordinarily, yodels and I've grown used to hearing all of these variations at pupping and moulting times. In those seasons, the seals' calls have obvious motives – preservation of personal space, protection of pups, rebuttal of lustful attention – but I've never heard an entire group howling in unison for no discernible purpose. The keening curls and coils, twines and twists, mirroring the dawn-pink Merseyside industrial smoke opposite, and though, to human ears, its tone might sound doleful and ghostly, after listening for half an hour, I'm zinging with joy.

In fact, if I had an mp3 of this instead of a disembodied voice telling me to progressively tense and relax all my muscles or breathe out for a count of eight, I feel sure I'd fall asleep in no time, regardless of how anxious I feel about Dad.

Bald crown, though it'll probably be covered by his cap. Clean shaven. Green eyes. Glasses. Thanks to the rain, he'll most likely have a red waterproof on – he used to wear it when walking in the Brecon Beacons and though the lining's torn and I bought him a new one, he still favours the old. Height? Five foot eight. Or he used to be. Age has curved his spine a little.

It's now past 11 p.m. When I first phoned the police, they recorded my description of Dad, along with details of his car, home and most recent movements. They gave me a unique reference number, said they'd prioritise the case because of his age and cognitive condition and promised to call me back within an hour with an update. When the hour was up, I waited for a further fidgety twenty minutes, then phoned again.

Kind smile. A lattice of laughter lines. A scar underscoring one eye from being hit by a cricket ball. Liver spots mottling the skin of one temple – a pattern I could instantly ID without the need for ExtractCompare software.

Nothing notable to report, the police told me. They phoned his landline and mobile, confirmed he isn't answering, and sent an officer to his home, confirmed he's not there. Found no record of admission to his local hospital. Planning to check traffic cameras for glimpses of his car.

Short fingers. Broad hands. Hands that coaxed me over streams, up trees. Hands that filled jars with earth or leaves for my worms and caterpillars. Hands that threw darts to save bagged goldfish from the fair, then dug a pond to home them.

So now, back to waiting. Haven't eaten so force myself to swallow toast and tea. Do an uncharacteristic deep clean of the kitchen. Hooper's sleeping in front of the wood burner and I lie there after midnight for a while, head against his warm belly, trying to replicate his slow, peaceful breathing.

Soft voice. Vowels still shortened by his Northern childhood. A voice that storied me to sleep with Ransome, Milne and Blyton. A voice that I might never—

My phone rings again. I lunge, knock it from table to tiled floor, scrabble around, yank it towards my ear. 'Yes? Yes? Hello?'

His car's been spotted by a camera.

The seals sing me onwards as I tackle the short distance between Middle Eye and Hilbre. Their vocals have quietened the chatter in my mind and my earlier worries about sinking in quicksand seem a bit overdramatic. This is rougher walking now, across barnacled and seaweed-slick rocks, dog whelks, winkles, and shards of glass that have yet to be sea-smoothed.

Hilbre, the second syllable of its name pronounced like the French soft cheese, is the largest of the islands at about eleven acres. I access it via a path at the southern end, passing a heron posing bonily on a rock like a Size Zero, and furrowed sandstone cliffs, richly gold and red in the morning sun. An information board with branding by Wirral Council, which owns Hilbre, details its credentials as a Local Nature Reserve.

Though I'm today's first arrival, there are signs of human habitation – a clutch of buildings, fences, a gravel track – which make for a less remote feel than was the case on Middle Eye. While the island's ranger is no longer resident – I came across an online article claiming that Wirral Council couldn't find anyone willing to live on site without mains electricity and running water – a couple of privately owned cottages are occasionally occupied. Another bungalow functions as Hilbre Bird Observatory: the Dee Estuary is recognised as an internationally significant site for waders – migratory and overwintering birds in particular – thanks to the muddy abundance of shellfish and other marine invertebrates. As well as listing species seen and sometimes catching and ringing birds, volunteers have also maintained records of sandbank seal numbers since the mid-twentieth century.

The building at which I linger longest is the old Telegraph Station, the door of which is locked but through whose curved, panoramic window I peer to check out what's inside. From what

I can tell, it's the island's interpretative centre, with a micro-exhibition about Hilbre's history and wildlife. I also spy a notice advertising monthly Seal Days throughout the spring and summer when a telescope offers close-up seal views and 'All being well and weather permitting, the Friends of Hilbre volunteers serve refreshments.'

I made contact with the Friends of Hilbre, a membership organisation that promotes the conservation of the nature reserve, before travelling to the Wirral, and managed to chat with David Gregson, the vice-chair. Though he was absorbed in looking after his very poorly mother, he still found time to enthuse about the sandbank seals and share his concerns about the challenges they're facing. Disturbance caused by boats from local sailing clubs seems to be especially problematic. 'We've written to them all about it but none of them replied,' he said sadly, highlighting the annual Commodore's Cruise 'when lots and lots of boats go out, land on the sandbank and have barbecues' as the occasion when the seals are most affected. 'It's quite a touchy issue.'

Leaving the interpretive centre, I saunter on towards the northernmost point of the island and the ruin of a lifeboat station. Here, I get settled on a rock and tune in once more to the seals' polyphonic song. It no longer seems so layered and resonant – many of the voices have fallen away. The tide has turned and some of the seals are now preoccupied with shifting higher up the shrinking sandbank, while others have chosen to slide off and swim. I'm planning to stay on Hilbre over the high tide period to observe the seal activity and it won't be long before water starts advancing up the lifeboat slipway.

'Peaceful, yeah?'

I do a clichéd flinch of surprise. A lanky man with a long stride and shin-high Doc Martens is moving at pace towards me. When he comes to a wide-legged halt next to my rock, I scramble to my feet, feeling suddenly vulnerable.

'Bonus not to hear any traffic.'

He's here to enjoy the calm and quiet of the island, nothing more sinister than that, I tell myself as his gaze roves over waves, rocks and ruin. Now that the seals are subsiding into silence, though, my anxiety blather is starting to fill the void again, questioning whether I should maroon myself here over the high tide after all.

Dark eyes veer back from sea to me.

Quick, visceral decision. Pack away binoculars, water bottle, map.

'You off?'

The strap of my binos gets trapped in my rucksack zip. I feel his eyes on me as I tug and fumble.

'Yep. Been here ages.' Sort the zip later. Throw rucksack over shoulder. 'Got to get back.'

'Well, have a good one, yeah?' Thin grin. 'See you around.'

Those last three words – menacing intent or a throwaway goodbye? As I try to resist turning round to check on his next move, the words of crime writer Ann Cleeves are spiralling through my catastrophising mind. Cleeves was resident on Hilbre in the late 1970s when her husband was employed as the island warden and she started writing her first novel while living here.

'It's a wild, wonderful, lonely location,' she said. 'The perfect place for a murder.'[1]

When the police phoned to say that Dad's car had been seen, no more than a tiny seedling of relief pushed through the crack in my panic. His car was spotted on a road he'd never usually use, some twenty miles from his home. How do I know he was the driver? Could his car have been stolen? Has he been mugged? Abducted? How can I know he's okay?

All I can do is wait again and I'm too wired even to try cwtching Hooper in front of the fire. I pace between kitchen and lounge, up and down the stairs, straighten towels on rails and pictures on walls, make more tea, leave the mugs on shelves and forget to drink it. If only Dad could be satellite tracked like a seal, a tag tenderly attached to the back of his neck and facts on his where-abouts and welfare dispatched to my own personal receiving station.

The next time the phone rings, I don't make a sudden lunge. Three seconds. Four. Five. Scared to hear what the police will tell me. Yet also desperate to know.

'Hello? Sue?'

Adrenaline spikes like a spring tide. 'Dad! Are you all right? Where are you?'

'Back home. And of course I'm all right.'

'*Of course*? What d'you mean of course?' Hooper raises his head and thumps his tail against the rug, as if exhorting me not to rage at Dad for sounding so blasé. 'Where've you been? What have you been *doing*?'

His explanation's long and rather garbled. Saw the podiatrist – rush hour when he left – raining. Roadworks – a diversion that took him on a route he didn't recognise. Lost the diversion signs. Lost his bearings. And then it got dark and his confusion intensified.

'Six hours, though – you've been driving around for six hours—'

'It can't have been that long.'

'It's one in the morning.'

'Is it really?'

I've got not a droplet of energy left in me, sag onto the settee. 'So did the police find you?'

'The who?'

'The police. I rang them – they were looking. . . How did you get home?'

Long pause. I try to resist the urge to bombard him with more questions. He'll only get more muddled – I need to let him take his time.

'I stopped at a big shop.'

'What, like a supermarket? Where?'

'Never mind where.' Another pause. 'I saw someone by his car unloading his whaddyacallit – that thing on wheels you put your food in. Told him where I was trying to get to and he said he'd get me back home.'

'You mean you latched on to some random man – told him your address, followed him in your car? You could have been—'

But he wasn't. He's safe and, as far as I can tell from this distance, well.

'Have a snack, Dad, and make some tea. Then get yourself to bed. I'll be over to see you first thing tomorrow.'

Now that I'm back on Middle Eye and the wide-striding man hasn't attempted to follow me from Hilbre, I sit cross-legged on the grass and try to make sense of my reaction. I'd never usually be spooked in this way – I often encounter solo male walkers on the coast path at Aber Felin but never feel the need to flee from them. Since the night when Dad went AWOL, though, my anxiety's skyrocketed – it's clustered around his well-being and living situation of course but has started to corrode many other aspects of my life, sleep included, now too.

'How about over here?' A man is guiding his female companion, her once-pink ballet pumps glooped in estuary mud, across the grass.

'Aw, cheers, Tim,' she says as he spreads out a blanket a few metres away from me. 'I'm well sorted now.'

To sort her further, he proceeds to give her a lesson on how to use his binoculars. 'D'you see all those little black dots in the water?'

'Dunno. . . Wait a sec. . . Yeah, think so.'

'They're seals.'

'Seals! You're joking me! Wow!' She beams at him like he's personally conjured them up for her himself.

In the weeks since the Missing Person incident, I've tried to put some contingencies in place to prevent something similar happening again. I've spoken to Dad's podiatrist and dentist to ensure he's only given morning appointments to remove the risk of his having to drive anywhere in the dark. I've stuck a laminated notice above the hook where he keeps his car keys reminding him always to take his mobile with him, and instructions on how to operate it are tucked inside his wallet. I've upped my phone calls to him from once to twice daily and the neighbour who checked to see if his car was in the drive now has a set of his house keys for emergency access. Any more radical changes to his routine and lifestyle, however, he's firmly resisting.

Only about twenty seals are still on the sandbank, firmly resisting the incoming tide. Closer to Middle Eye, the sickle beaks of a conclave of curlews are grubbing for lugworms and molluscs. Meanwhile, a bald man with a neat moustache, the bottom half of his cords miraculously free of splotches of mud, has arrived on the island. He precisely sets out on the grass a sketchpad, a mini-tin of paints and an ancient paracetamol bottle full of water for rinsing his brush.

Right up until the hour before I left for the Wirral, I dithered over whether I should come. This is supposed to be the first of many spring and summer seal site visits but because of Dad's circumstances, travelling's started to feel a bit precarious. I finally convinced myself, though, that I live several hours away from him anyway, so apart from the reassurance and sense of familiarity he may feel from my being in Pembrokeshire, there's no real difference in practical terms if I need to make an urgent dash to see him.

The man with the paints is trying to mix an appropriate shade of red to capture Middle Eye's and Hilbre's sandstone cliffs. Rather more frenetically, an RNLI lifeguard is zooming across the sand on a buggy, having spotted a family trying to take a potentially treacherous shortcut to Hilbre rather than follow the safe, recommended route. After a lot of gesticulating, he manages to stop them proceeding further.

The remaining few seals on the sandbank are now completely silent. In addition to continuing my seal journey as serenely as possible in the context of my worries about Dad, I'm also hoping, as a result of this morning's glorious chorus, to find out more about seal vocalisations as I travel further north in the weeks ahead.

'Tim, this place is awesome!' The woman near me on the picnic blanket is still watching the swimming seals through binoculars and still beaming.

'Yeah,' he says. 'I'm not going to lie – this is our Uluru.'

9

Action Man

At 6 a.m., bowl of porridge in hand, I amble from my lighthouse cottage to the bright red-painted foghorn. Its gape is mirrored by the red bill of the chough perched on top, constantly calling – *chee-ow! chee-ow!* Having picked my way down onto the rocks below, doing my best to avoid the delicate straggles of sea ivory lichen, I sit and wait for my porridge to get cool enough to eat. A scurry of ringed plovers crests an adjacent rock, while next to my left thigh, a bumblebee comes in to land on a pink thrift flower, causing it to make a low bow. The jumble of rocks stretches another twenty metres or so to the sea's edge where each is topped by a sleeping grey seal like a wedge of candied peel on a cupcake.

Though I've never been here before this week, it feels, in some ways, very familiar – the late spring flora and fauna, as well as the topography, are reminiscent of many stretches of Britain's western seaboard, including Aber Felin. Yet my purse contains notes and coins from an unfamiliar currency, and on the radio, stories pertaining to the UK are referred to as 'international news'. Famed for its financial and technological sectors and named, so it's claimed, after Manannán, a Celtic sea god, the island to which I've travelled slants between Northern Ireland and England as if it's in italics. A four-hour boat journey with the Steam Packet

Company brought me to this latest destination on my grey seal journey – the Isle of Man.

The cottage where I'm staying lies at the southern extremity of the island, on the tip of the Langness Peninsula. Adjacent to the automated lighthouse, it was owned, until fairly recently, together with two other keepers' cottages, by ex-*Top Gear* presenter Jeremy Clarkson, whose on-screen bombast seems utterly at odds with the peace and beauty of the location. The sea's so flat today that a perfect, unrippled reflection flies beneath its cormorant, while the sky's so clear that the sun heightens the colour of everything it touches. The beak of an oystercatcher glistens vivid orange after dabbling in dewy thrift and grass and the feathers of the chough on the foghorn shine glossy black.

The sun is casting a glow on the seals' bellies too as they snooze on their sides on the rocks. It's even harder than usual to resist anthropomorphising as all nineteen of them seem to be luxuriating in the early morning warmth. One particularly languorous cow has a long stretch, rubs her foreflipper over her face, then throws her head back beyond her rock's edge to soak up more rays.

A brief avian skirmish disrupts the tranquillity. When a herring gull enters the airspace above the foghorn, another chough sweeps in and swoops back and forth to drive it away. The gull offers only half-hearted resistance before flying off again and the triumphant second chough joins the first on top of the foghorn's trumpet. As I start to tuck in to my porridge, chough number one resumes its insistent calling, wings lifting with the intensity of its effort. Is this an alarm call? The gull's been sent on its way, so where's the perceived danger?

Another seal, a young cow, has swum into view but she's too far away to be bothering the choughs. Instead, she's pestering the sunbathers, trying to heave herself up onto a rock that's already occupied. The cow she rouses hisses, moans and lunges at the new arrival, who slides backwards into the water, swims in a chastened circle, then returns to try to gain purchase on a different rock. Again, its occupant bays with irritation at being disturbed and refuses to make space for her. Next time, she holds back a little, extends her neck and sniffs at the rock and the seal that's topping it instead of trying to muscle her way up. Her presence still isn't welcome, though, and several others join in with the vocals to tell her so.

Apart from this minor grouching to preserve their personal basking spots, this gathering of seals is silent, with none of the choric howling I experienced on the Wirral. There's none of the hormonally heightened aggravation that's characteristic of a breeding colony either and much as I love seeing libidinous bulls and edgily protective cows, my anxious mind needs this relaxed spring haul-out. The boat journey, plus the Manx government's online insistence that the island 'is not, and has never been, part of the United Kingdom',[1] have increased both my feeling of distance from Dad and my concern about how easily I'd make an emergency return. That said, things seem calm and stable with him at the moment and, even though every time I phone, he initially thinks I'm still in Pembrokeshire ('How are things out west today?') he soon reorientates himself to my new location ('Did I ever tell you about the time I went to the Isle of Man? Not long after the war when we lived in Chorley. Dreadful crossing').

While my time here is, of necessity, limited, it's the biggest island, at thirty-three miles long and thirteen miles wide, that I've thus far visited on my journey, with an extensive, potentially seal-rich coastline to discover. Fortunately, I've been able to make contact with Lara Howe, marine officer with Manx Wildlife Trust and the island's leading seal expert, to seek advice about narrowing my area of exploration. Though we hoped to meet at some point during my stay, she's ended up having to work off-island so we've connected via phone instead. As well as telling me about the seal monitoring work that's undertaken here, she's also sent me some reports and hotspot maps, all of which have confirmed that the rocks I'm currently gazing at support one of the island's most important haul-outs. As is the case with the Wirral sandbank seals, however, numbers on the Langness Peninsula tail off in the autumn and winter, indicating that most leave for breeding sites elsewhere.

Lara explained that the first monitoring of consequence took the form of a monthly boat survey of the entire island in 2006–7 to gauge numbers, map haul-out and pupping sites, and record instances of disturbance. Ten years later, another all-island survey took place, with Manx Wildlife Trust scrutinising the whole coast over several October days to count adults, pups and juveniles. In the intervening years, annual pup monitoring of the island's most notable breeding site, the Calf of Man, was carried out.

This might seem like relatively small-scale research compared with other locations I'm familiar with in Cornwall and West Wales, yet Lara assured me it has little to do with lack of intent and everything to do with the availability of funding. 'We've got over ten years of data – I've managed to keep it going by eking out bits of money. Just wish I could keep it going now on a more formal basis.'

I make my porridge last for a full forty minutes, by which time the young cow's vying with the tide to claim available rock space and those in possession are trying not to get washed off. Through binoculars, I spot another gathering of seals on another wodge of rock where the peninsula tip tapers west and decide to wander across to watch them. As soon as I leave the area below the foghorn, the chiding chough falls silent, suggesting that though I was sitting quietly and moving spoon between mouth and bowl slowly and mindfully, I was, for some reason, deemed to be a threat.

After dropping off crockery and cutlery at the cottage, I become aware that I'm following the footpath to which Jeremy Clarkson controversially blocked access over a decade ago. He took exception to its passing so close to his home that, according to his now · ex-wife, people could 'come up to him at close range and take photographs even when he's sat eating a pork pie'.[2] A number of locals, whom he dubbed 'very unpleasant militant dog walkers',[3] proceeded to set up a campaign group, Public Rights of Way Langness, with the felicitous acronym of PROWL. They argued that, although it wasn't officially a public footpath, the landowner had historically granted permissive access, and since people had used it for well over the legally required twenty-one years without interruption, a right of way had been established. The dispute grumbled on for over half a decade, reaching High Court level, and while Clarkson was ultimately ordered to reopen the path, his threat to give up the lighthouse cottages 'and go and sit under the arches of Waterloo Bridge'[4] if he lost the legal battle turned out to be an empty one.

Clarkson no longer has any association with the lighthouse site and his ex-wife is now the sole owner. Yet even today, a number of years on from the resolution of the dispute, I find a few bossy signs fixed to the fence alongside the footpath. All Dogs Must Be Kept On A Lead From This Point. No Exceptions. Non Compliance Will Lead To Prosecution. Hooper's come with me to the Isle of Man and though I'm more than happy to walk him on-lead, especially as Highland cattle have been introduced to the peninsula to graze the vegetation and extend the habitat of the lesser mottled grasshopper, a rare species that's said not to exist anywhere else in the British Isles, I can't help but feel that more benign signage would be welcome.

Eleven cow seals are sprawled on the outcrop at which I'm now looking – the absence of bulls fits with what I read in Lara's most recent report about the 'prevalent female majority in the observable seal population'. They're all quite fidgety, lifting heads and flippers whenever the sea creeps further up the rock, shifting position, pursuing the high and dry. Though they pay no attention to the noise of the plane taking off from nearby Ronaldsway Airport, my back and shoulders tighten in sympathetic response to the thrust of its climb. Moments later, a lark rises to my right, trilling its territorial song, its high-into-the-sky ascent even more impressive than the plane's.

Sitting on a grassy mound blued with scribbles of spring squill, I watch seal after seal yield to the tide till only one remains.

'The Sound is the primo spot to see them. If you don't see a seal in the Sound, you've gone with your eyes shut!'

Having enjoyed a day on the Langness Peninsula in close proximity to my cottage, I spend the next day further afield, acting on Lara's advice to visit the site of greatest seal significance. The Calf of Man is a small island and nature reserve, separated from the Isle of Man's southwest coast by Calf Sound, a thin strait churning with rip tides. The even smaller Kitterland lies in the centre of the strait and the whole area hosts both an autumn breeding colony and a year-round haul-out.

My mood, as I board the little boat that will take me across the Sound and circumnavigate the Calf, is a peculiar mix of anticipation and melancholy. It happens to be my birthday and it's the first year I haven't received a card or an early morning phone call from Dad. Hooper, meanwhile, a veteran of seal-spotting boat trips, is the opposite of melancholic. In contrast to the peremptory Langness dog-handling advice, Jason, the on-board guide, urges me to let him off the lead so he can wander freely around the boat. After greeting the other passengers – a man with three cameras round his neck, two women on a mission to spot dolphins, a man with a deluded 'CAPTAIN' printed on the front of his baseball cap – he finds a sunny spot on deck and flops down on his belly with a contented sigh.

As we pootle out of Port Erin and follow the coast south to the Sound, Jason chats about his Manx ancestry – a grandfather who ran the island's first animal circus and a grandmother who sold cigarettes on the trams – and passes round a laminated list of the species, in addition to grey seals, that we should look out for today. The dolphin devotees are delighted to find three species listed – bottlenose, Risso's and short-beaked common – and we're at the start of basking shark season too. In late spring and summer, the Isle of Man's plankton-rich waters attract one of the world's largest concentrations of this immense, endangered shark, though, worryingly, over the past few years, sightings have fallen.

Just like yesterday, the sea is serene and the sky is clear. The silhouette of Ireland's Mountains of Mourne undulates above the western horizon and I can hear a range of subtle sounds that would otherwise be drowned out by wind and waves. The cooing of male eiders, like my mother's exclamation of embarrassed surprise if ever Dad shared a risqué joke. And, as soon as we reach the craggy islet of Kitterland, the huff of a bull seal preparing to dive.

As Lara promised, there are seals aplenty in this area. Drifting down the west coast of the Calf of Man, we pass rock after rock that's occupied by a dozing adult or juvenile. A few raise their heads but otherwise, they pay us no attention. It would be so easy for a boat to flush them all into the sea and I'm relieved that ours is keeping an appropriate distance. When I manage to drag my eyes away from the seals to the cliffs above, I enjoy the sight of some nesting guillemots, a couple of hooded crows and splashes of white sea campions and scurvy grass. There are one or two seals in the water too, including a perky juvenile who watches our progression with curiosity, dives as we pass, then reappears a few metres ahead to follow us with her eyes again.

This sequence of behaviour, repeated umpteen times over, lures the man with the three cameras to the side of the boat. 'C'mon, I want the classic Nessie shot with your head above the water,' he says. 'C'mon, show us your whiskers.'

Though Hooper would usually be equally animated by the scent and sight of a seal, he's in snooze-in-the-sun mode right now. I've sometimes speculated whether his customary fascination is born of some kind of distant kin recognition as seals and dogs are in the same taxonomic suborder, Caniformia. However, the pinniped family split from other caniforms some fifty million years ago so this notion is admittedly a bit fanciful.

There's ongoing scientific debate, too, as to whether seals share closer ancestry with bears or mustelids, the family that includes otters. A relationship with the latter seems to be apparent in the shape of a skeleton from around twenty-one to twenty-four million years ago, during the Miocene epoch, that was discovered in the Canadian Arctic in 2007. Given the genus name *Puijila* after an Inuktitut word referring to a young seal, the skeleton offers morphological evidence of one stage in seals' transition from terrestrial carnivores with legs to flippered marine mammals. *Puijila* was probably proficient at both walking on land and swimming as, though its flattened toe bones signify webbed feet, they hadn't yet developed into flippers. With its long tail and short, sturdy legs, it gives the impression of being very otter-like, though its skull is said to be distinctively sealy.

Another Miocene-era ancestor of seals, *Enaliarctos*, which frequented the waters off what's now the northwest coast of the USA, had already evolved flippers, although the existence of carnassial teeth for slicing and shearing – common in carnivores yet absent in today's seals – suggests it would return to shore with prey items to process them. The long claws and flexible 'fingers' of its foreflippers indicate that it would have been able to grasp that prey, however, just as grey and other phocid seals now use their front flippers to grip the fish they've captured in their jaws, then tug it downwards to tear it. Seals' retention of paw-like foreflippers aside, the principal legacy of their terrestrial past is, of course, their practice of hauling out on land to breed.

From the dark, abyssal zone of pinniped prehistory, my attention swims back up to the sunlit surface of the present. Now that I'm sailing round the Calf of Man, I can match the maps and other details in the reports that Lara sent me to actual locations. Usually, in the pupping season, a succession of volunteers stays on the island for a week or two at a time to carry out surveys of twelve key sites. More recently, though, a couple of researchers were able to reside on the Calf for the full six weeks of the season and conduct a more comprehensive survey on a daily basis, thanks to a modicum of funding that became available from a Manx Wildlife Trust campaign. 'That went towards their flights and subsistence,' Lara told me. 'Then after they finished on the Calf, I put them up at my house for two weeks so they could write the report – the cost of accommodation here's ridiculous.'

The researchers' time on the Calf coincided with the arrival of Storms Ophelia and Brian and their report acknowledged that while there had been a steady increase in pup numbers from the twenty-six that were recorded when surveying got seriously underway in 2009 to the peak of eighty-four in 2016, survival rates in the year of the storms were, inevitably, lower.

'Cath! Look!' One of the women on board leaps to her feet.

'What?'

'Dolphin!'

'Where?' Her friend jumps up too.

'There! By that rock. . . Oh.'

'Oh what?'

'It's just another seal.' She sinks back down again, gesturing towards the snout of a bottling bull, mistaken for a fin.

Apart from climate change giving rise to more storm activity, I asked Lara what she considered to be the biggest challenges facing seals around the Isle of Man. As the Manx government is trying to attract extra people to live on the island, she acknowledged that human disturbance incidents could increase, while the tourist industry is contributing to more leisure craft usage and boat trips, especially around the Calf. Compared with the late nineteenth and early twentieth centuries, however, when Douglas, the capital, was a fashionable resort, described in the title of a 1904 pamphlet as 'the Naples of the North', with a thousand boarding houses and dance halls lining its two-mile promenade, tourist numbers are currently manageable.

'What about shootings and the attitude of fishermen?' I went on to ask.

'Well, no dead stranding has ever indicated a shooting – but then we don't have fin fishery. It's scallops, queenies, whelk, crab . . . I suspect our fishermen don't love seals but they don't seem to hate them either.'

My journey to the Isle of Man from the port of Heysham in Lancashire got me thinking about the impact of wind farm developments on seal populations too. En route, we cruised close to numerous rows of sedately rotating turbines, comprising the biggest group of operational offshore wind farms in the world. Having been alerted to the potential dangers of the tidal energy unit in Ramsey Sound, I'm aware that renewable energy development can pose welfare threats to marine wildlife, and there have been indications that the booms of hydraulic hammers during the construction of wind turbine foundations – one strike per second, repeated up to five thousand times in some instances – exceed seals' hearing damage thresholds. In addition, grey seals vocalise beneath the surface much more habitually than was once thought and it's likely that the extreme noise, which can transmit through water for many miles, masks their communications.

There have, though, also been signs that, once installed, wind farms can benefit seals. A study analysing the North Sea-based movements of a number of animals, through satellite tracking that originated in the UK and the Netherlands, concluded that some were drawn to forage around offshore wind turbines, along with underwater pipelines, demonstrating that they must act as artificial reefs. Fish are attracted to the crustaceans that cluster on the man-made structures and this, in turn, seduces seals. The first of the gigantic wind farms that we passed on the way to the Isle of Man is located less than ten miles from Cumbria's Walney Island near Barrow-in-Furness, where a burgeoning seal haul-out and nascent breeding colony is located. Though further research is needed, it seems possible that an aggregation of prey in and around the wind farm could be contributing to Walney Island's steady increase in numbers.

'Dolphin!' shouts one of the women again, causing Hooper to wake and spring to his feet, then look quizzically at me for an explanation of why he needs to be on high alert. This time, it's definitely a fin and not a seal snout that's been seen, though from its triangular shape and small size, I'd say it belongs to a harbour porpoise.

We're heading up the east coast of the Calf of Man now and there are no seals in view at all – they're favouring the western and southern sides, taking advantage of the afternoon sun. Instead, we spy signs of human presence – stone walls, the seasonal

wardens' accommodation – as well as some decoy puffins positioned on the cliffs in the hope that they'll attract breeding pairs.

'Puffins always used to nest here but then the longtails got them
– the R-A-T-S,' says Jason, spelling the word out. 'And if anyone
says that word on my boat they'll be thrown off.' He explains that
an r-a-t eradication programme has been underway on the Calf
for some years and though puffins haven't yet returned to breed,
numbers of nesting Manx shearwaters are now rising. It pleases
me to realise that this extraordinary bird, familiar to me from
Skomer, is recolonising the very place from which, thanks to its
historical abundance, its name originated.

I glimpse another iconic Manx animal too – the Calf is home
to a flock of rare Loaghtan sheep, famed for their dark fleece and
the two, four, or sometimes even six horns of both the ewes and
rams. We pass beneath a ewe, grazing nonchalantly on the narrowest
of ledges while her lamb bleats anxiously from the cliff top.

The tide race pulls us through Calf Sound where a corner of
Kitterland is now occupied by a throng of over twenty seals, their
patterned fur transforming the rock into a marbled countertop. I
quickly assess that there are five large bulls in the haul-out and
try to work out if one of them could be the seal for whom Lara
instructed me to keep an eye out, Mr Velvet. 'He's humungous –
massive neck, rolls and rolls of fat – and very scarred. He gets
the primo sites at pupping time and access to a good number of
ladies. He's well ugly – I don't know what the ladies see in him
– but he's got to be one of my favourites.'

As well as recognising Mr Velvet by eye, she revealed that she
maintains a photographic ID catalogue, without the benefit of
ExtractCompare, which now includes over four hundred individuals. The majority of those who give birth on the Calf are
regulars – in the Storm Ophelia-scarred breeding season, forty-
nine cows out of sixty-six were confirmed to have pupped here
before. 'We don't know yet if the new individuals are seals who
were born here who've come back to have pups themselves or
seals who are completely new to the island,' she says, referring

to the fact that some cows and, to a lesser extent, bulls, are philopatric, returning to their own birthplace to breed. 'I'd love to get some tags and satellite track them but you need serious money for that.'

As we steam back towards Port Erin, the man with the cameras battles to capture the full wingspan of a gannet while the dolphin addicts do a last-ditch panoramic scan of the sea. The Mountains of Mourne are in view again, outlined against the sky, reminding me both of Ireland's proximity and the Isle of Man's pivotal, mid-Irish Sea location. The number of grey seals that populates the entire Irish Sea region is estimated to be between five and seven thousand and, over the past few decades, some research has been conducted into their journeys between southeast Ireland and southwest Wales, largely through photo ID work. It can be assumed that they also move freely between the Isle of Man, England and Scotland, though the extent of these movements is little understood.

Once we reach Port Erin's horseshoe-shaped bay, Jason deals with all the palaver of mooring, while Hooper stretches and sniffs the air. His mind is set on exploring the promenade now, scouting out dropped chips and dripped ice cream.

My mind, however, remains anchored in the Irish Sea. Lara's sundry allusions to financial constraints during our chat reminded me that while seals of course recognise no geopolitical boundaries, the government bodies, agencies and conservation NGOs in the countries that surround, and on the island that's in the middle of, the Irish Sea are likely to have quite different seal-related budget allocations, research concerns and management priorities. And these disparities are on a far larger scale, with perhaps far wider implications, than the contrast in monitoring methods that I came across on Ramsey and Skomer.

The new moon's low in the sky, an orange shred of thin-cut marmalade. I'm spending my birthday evening on the rocks below the foghorn at Langness in the company of nine seals, a pair of oystercatchers and the chough that chastised me at yesterday's breakfast.

There was a *Groundhog Day* flavour to my earlier phone call with Dad ('How are things out west, then? . . . Ah, that's right, you're on the whatchamacallit – the Isle of Man. I went there once – did I tell you? After the war when we lived in Chorley.') But he's had a calm, crisis-free day and my usual pre-call anxiety has receded enough for me to enjoy a picnic of crackers, tomatoes, apples, crisps and cheese.

I'm about to start on a slice of bannog, a traditional Manx fruit bread, my birthday cake stand-in, when the second chough appears. Not, though, from the sky or one of the thrift-tufted crags, but from inside the trumpet of the foghorn. As it joins its mate on top, I finally grasp the reason for all the alarm calls and why the herring gull was sent on its way too – there must be a nest in there. Fortunately, the foghorn's been decommissioned and no longer blasts any warnings, so the chough chicks will be able to grow in peace.

I stay outside long beyond sundown, leaning back on my elbows to take in the density of stars. I hear the lusty piping of a single oystercatcher and the softer peeping of the pair that's settled among the rocks just a short distance away.

A seal at the shoreline harrumphs and snorts while another two howl a grumpy duet.

And the waves, like my father, repeat themselves over and over.

10

Learning to Speak Seal

'My Name's Doolittle And I Can Sing To The Animals. Mum Turns Sea Creature Sounds Into Music.'[1]

I was first alerted to the work of Dr Emily Doolittle by an article's corny headline in the Scottish *Sunday Mail*. After a brief return trip to Wales, I've travelled north again, to Glasgow, for my annual spring writing residency with the British Animal Studies Network at the University of Strathclyde, and while this is always one of the highlights of my year, it's the most landlocked and city-based I've been for a long time. I'm hoping that, in between my residency commitments, meeting Emily and learning about her work as a composer and one of the world's few zoomusicologists will help keep my seal journey on track even though I have no sight or sound of the sea.

Coined by French composer François-Bernard Mâche in *Music, Myth and Nature, or The Dolphins of Arion*, first published in an English translation in 1992, zoomusicology is the study of the musical features of nonhuman animal sounds. Though I thought I'd got to know Glasgow fairly well over the past few years, it seems I've never before made it to one of its hippest neighbourhoods and I read a few key extracts from Mâche's book in a characterful café as I wait for Emily to arrive. 'If it turns out that music is a widespread phenomenon in several living species apart from man,' he writes, 'this will very much call into question the definition of music, and more widely that of man and his culture, as well as the idea we have of the animal itself.'

I'm in the process of selecting a sensible slice of sourdough toast over a chocolate and Guinness cupcake when Emily appears, eyes warm behind her oval glasses and smiling broadly. Like all the other customers, whether eat-in or takeaway, she enjoys a chat with the barista, after which she sits opposite me, rearranging the cushions on her portion of the wall-length bench for maximum comfort. Originally from Nova Scotia, Canada, she and her young family are now based in Glasgow, where she's a Research Fellow and Lecturer in Composition at the Royal Conservatoire of Scotland, and she gives the impression of being thoroughly at home here.

I resist the urge to open with a glib comment about the aptness of her surname – she must be seriously weary of this by now – and ask, instead, about the circumstances and influences that led to her becoming a zoomusicologist.

'Well, I was always interested in animals and music but I didn't connect them for a while,' she says. Though she participated in several workshops and creative collaborations under the guidance of composer and acoustic ecologist Murray Schafer in an Ontarian forest, her passion for music and nature interaction was, rather improbably, only fully aroused in a city setting in the Netherlands. Between completing her Masters in composition and embarking on her PhD, she spent two years at the Koninklijk Conservatorium in The Hague. 'Soon after I moved there, at around four in the morning, I heard a bird singing outside my window – it was so beautiful. I'd never heard it before and I found out it was a black-bird.' She became fascinated by the realisation that certain sections sounded like human music 'with little scales and arpeggios' though it was evident that blackbirds jump from one motif to another without any sense of connection, in contrast to the linking material that's a feature of the classical music tradition. When the original blackbird she heard outside her window went away and 'a really terrible one came in, a young one' that only sang two or three different phrases, she was nevertheless able to begin working on composing *night black bird song*, thanks to having 'a friend with a great blackbird – I spent time on her porch transcribing that!'

Never did she imagine, as she considered human and blackbird ways of arranging the same motifs, that she'd still be exploring the relationship between human music and animal songs in her compositions some two-and-a-half decades later. Over the grind, hiss and gurgle of the coffee machine, she reveals that, in addition to continuing to pursue research into, and produce music that engages with, the songs of a range of avian species, she's focused, over the years, on various mammals too. 'I'm interested in all those animals who have vocal learning, who learn from others of their species how to vocalise or sing. So not just songbirds, but other creatures such as whales and bats – and seals.'

I sit forward, resting my elbows on the tiny table among teapot, flat white, notebooks, pens and plates, to better absorb what she's going to say. Prior to her current Research Fellow role, she tells me that she enjoyed several previous visits to Scotland, including a three-month residency to study Gaelic folklore and learn the fiddle. Her first encounters with seals took place on the Isle of Bute in the Firth of Clyde, which can be reached in just ninety minutes from Glasgow. Both grey and common seals haul out in a bay there and Emily has noted their varying responses to human music. 'I hitchhiked there with a friend – we sang and played the fiddle to a group of ten or fifteen greys who seemed to be quite responsive. Another time, I played the fiddle to a group of common seals and they took no notice at all. With greys, it feels like the curiosity is mutual.'

'As well as the contrast between the two species, have you noticed if grey seals respond in different ways to different instruments?' I ask.

As she considers this, she rests one leg on her opposite knee and toys with her shoes' turquoise laces. 'When I did some work with schoolkids on the Isle of Bute and split them into groups, one group sang Gaelic songs, another had a French horn and the grey seals howled right back. Overall, I'd say they seem to like the voice and higher-pitched instruments. But it's hard to tell – I've never taken a cello out there, for example!'

Her higher-pitched instruments comment reminds me of an article I read recently about the Cornish Seal Sanctuary hiring a flautist to play live classical music to the resident seals and rescued pups after a research student's project concluded that it calmed them. The accompanying video featured the flautist in skirt and green wellies playing Bach at the poolside where an apparently impassive Snoopy was hauled out in her habitual spot. It looked like business as usual for the rabble of rehabbing pups too, yet the curator of the sanctuary insisted that feeding times were much less frenzied than usual and the blind seals, in particular, were soothed by the flute. 'We will be getting some speakers set up on site so we can play all our residents classical music in the mornings,' she concluded, 'as we know they find it relaxing.'[2]

Emily's laptop has emerged from her rucksack to join the clutter on the table so she can show me the score of the first of her seal-inspired compositions. Commissioned in 2011, *Seal Songs* is a thirty-minute piece for children's choir, chamber ensemble and narrator and it draws on one strand of the Gaelic folklore she originally came to Scotland to study: shapeshifting selkie myths. It's a long time since I tried to read music and the score contains a number of unusual squiggles, loops and curves that never appeared on the treble clef stave when I was in my school orchestra. These represent the acoustic landscape that the choir is required to create: cries of curlews and gulls, the crescendo of a storm, seals keening. The narrative moves through the usual stages of a selkie tale, from a fisherman's theft of a sealskin to his selkie wife's flight back to the sea many years later when she finds it again, and I just about manage to follow the music as well as the text, despite the fact that the notes I translate into sound in my head are periodically accompanied by the rumbling of the subway trains passing directly beneath us.

Emily admits that *Seal Songs* remains human-centric, based on our interaction with, and impression of, seals, rather than accurately replicating their vocal behaviour. She comes closer to capturing their essence, she feels, in the recently premièred *Conversation*, which emerged from her working alongside Professor Vincent Janik, an animal acoustic communication expert, at the Sea Mammal Research Unit at the University of St Andrews. Written for soprano, and an ensemble of five wind and five string instruments, it also features percussion in the form of the evocative ocean drum.

The first part of Emily's research process involved spending time in the field, gathering recordings of grey seals from an estuary haul-out north of St Andrews. 'Sometimes it felt, from the recordings, that the seals were howling in a particular pattern – ascending or descending – and then another seal would jump in with the same kind of howl,' she observes. 'I wanted to think of the contour of each howl and layer similar ones on top of each other.' She also hoped to convey the spirit of spontaneity that characterises a seal chorus and instructions in the score give the musicians the freedom to choose when to join in with the seal sounds and whether to play certain sections once, twice or three times. 'There's partial improvisation – each performance will be flexibly different and unexpected melodies and harmonies emerge as the sounds intermingle.'

Recalling the complex sonic overlappings of the seals on the sandbank on the Wirral, I'm struck by how fitting this creative choice seems. I'm also struck by the fact that, while it's undeniably sealy, *Conversation* still retains a strong human dimension, structured as it is around the setting of a poem of the same name by Emily's compatriot, Eleonore Schönmaier. 'Are my thoughts so noisy they murmur / outside my body?' the poem begins, artfully expressing Emily's own experience of tuning in to grey seals, the flavour of their vocalisations often strangely akin to that of humans.

'The past / and its regret talks / back to me . . . a conversation fetched home.'

The repetitive crunch of footsteps on sand, sounded by the bow of a double bass. Bird calls from the woodwind. The susurration of the ocean drum. Glissando effect from the strings – glide and slide from note to note – to approximate the seals' howling. And, woven through the soundscape, the rather angtsy soprano persistently asking, 'Are my thoughts so noisy, so noisy. . .? Are my thoughts so noisy. . .?'

Emily has generously emailed me a recording of one of the first performances of *Conversation*, by the St Andrews New Music Ensemble, and I listen to it several times over, sprawled on my bed in a Glasgow guesthouse.

'Are my thoughts so noisy, so noisy. . .? Are my thoughts so noisy. . .?'

It's almost time to make my evening check-up-on-Dad call and though I'm my usual twitchy self as I anticipate what problems might have arisen since this morning, I'm also keen to talk to him about *Conversation*. If he were to hear it, he'd describe it as 'modern', as he claims all classical music is, post-Brahms, but he wouldn't dismiss it – he'd listen closely and reflect. Like the seals at the Cornish Seal Sanctuary, he finds music deeply soothing and it's the source of the most vivid of his remaining memories too.

'Did I ever tell you about the time I played the *Moonlight Sonata* on the piano to try and impress a girl I fancied?' he asks, his story interspersed with arpeggios of laughter. 'Had a few school friends round when I was about fifteen – there I was, plinking and plonking away, and all the while she was necking some other boy on the sofa behind my back!

'And did I ever tell you how I met your mother?'

Mum managed a record shop in industrial South Wales. According to family lore, her grandfather, who hailed from Oxford, was asked, early in the twentieth century, to go into partnership with W. R. Morris, of car manufacturing fame. In what must rate as one of the most epic misjudgements of all time, he pronounced that there was no future in motor cars and moved to Wales to open a bicycle shop instead. Later, he diversified into gramophones and radios, and the record shop, run by my mother in the late 1950s and 60s, was part of this enterprise.

When work transferred my shy, self-effacing father from Northern England to South Wales, the music in my mother's record shop offered a welcome familiarity. The awkward *Moonlight Sonata* episode hadn't diminished his enthusiasm for Beethoven and he sought the latest Mozart and Schubert recordings too. Before and after he entered the listening booth, my chattier, more worldly-wise mother dispensed extensive advice on different music genres as well as on numerous other aspects of existence. It wasn't long before *Sinatra Sings of Love and Things* and *Mink Jazz* by Peggy Lee were as integrated into Dad's LP collection as Mum was into his life.

Like a sandbank shrinking in size with the rising tide, his reef of more recent musical memories has dwindled. The occasion I'd most love him to recall took place in the summer after my finals. Having sat ten three-hour exams in a week, I'd moved back home from London to Wales to wait for the result, and was spending the time writing and arguing vehemently with my mother. I had two writing endeavours underway: the first, a play for the embryonic Something Permanent Theatre Company, and the second – a more lucrative means of escaping another summer of waitressing – formulaic romantic stories for women's magazines. Mum grudgingly appreciated the publication of 'Another Sizzling Summer Story by Susie Richardson!' but regularly and insistently informed me that since I'd never make a living from writing plays, I should 'stop fiddling around and do something more useful'. Dad kept well away from these confrontations, retreating to another room with a crossword and a troubled expression or deciding the dog was in immediate need of another walk.

On the day on which the envelope containing my degree result arrived, I happened to have a ticket booked for a concert. Rebel violinist Nigel Kennedy, whose music my mother refused to listen to 'until he spruces himself up a bit', was on his Monster Bash tour and I'd been looking forward to going for weeks. If the envelope contained disappointing news, though, I'd more than likely feel inclined to hide away instead.

I took it into the garden, pushed my way through the dense row of conifers by the drive to the earthy, twiggy space that was my childhood den. Sitting with my back against our neighbours' fence, I rootled around on the ground for the trees' pea-sized cones just like I always used to when I was anxious.

If I find five, I'll open it.

If I find ten.

Twenty.

When at last I persuaded myself to unstick the flap, mud was clumped under my fingernails and I smeared the astonishing line that told me I'd been awarded a starred first.

After sharing the news with Mum and Dad, I went up to my room and sat at my desk. I had a schmaltzy story to finish and my keyboard tippity-tapped over the sounds from Mum's videoed episode of *Emmerdale* in the lounge below.

The next time I looked up, it was to see Dad hovering in the doorway.

'D'you think there are any tickets left for tonight?' he said. 'I'd really like to come along.'

Whenever I remind him of the evening now – the July heat, the two of us sitting, then standing and clapping and whooping as the music swung between jazz, rock and classical – he frowns and shakes his head, as if he can't believe it happened, as if I'm talking about some other concert attended by some other daughter and father.

I've long since given up on *surely you remember, it was such a great night, you loved it, please try. . .*

Enough to know that he was whooping as much for me as for Nigel Kennedy's virtuosity.

That I was whooping for being able to go back to London to do a funded MA and write my play.

And that both of us were whooping as we'd snagged the last ticket and could relish the evening together.

With my writing residency with the British Animal Studies Network in Glasgow over for another year, I've decided, before I continue my journey north, to travel due west to the Isle of Bute to visit the bay where Emily first sang and fiddled to both grey and common seals. I didn't, however, expect my first ten minutes on the island to be spent in 'the most beautiful urinals in the world'.

Having disembarked the ferry in the main settlement of Rothesay and glimpsed its lawned and tuliped Winter Gardens, I've detoured into the public loos, an unprepossessing brick building on the pier. After using the Ladies, I'm a tad perplexed when an attendant hands me a leaflet and ushers me into the Gents. They were built in the late Victorian era, she explains, as I take in all the mosaic and marble, blue and gold ceramic tiles, and the elaborate, fountain-like urinal structure in the centre. It's not surprising that I was under-whelmed by the Ladies – they were only added in the 1990s and opened by Lucinda Lambton, author of *Temples of Convenience and Chambers of Delight*, an illustrated 'history of the lavatory'. Though the Victorians neglected to provide for women's needs, she describes the urinals as 'jewels in the sanitarian's crown'.

At around fifteen miles long by five miles wide, Bute's a manage-able size for a day visit and, as I head across the mildly hilly interior to Scalpsie Bay on the island's west coast, I give a bit more consideration to what I might do when I get there. My ruck-sack contains all the usual paraphernalia for a day of seal watching with one quirky exception. In addition to binoculars, camera, notebooks, waterproofs, sunscreen, sandwiches and water, I've also brought along my ancient descant recorder, unplayed since the age of thirteen. The mouthpiece still bears the grooves of my nervous chewing and I'm half-convinced that if I were to take it apart and shake it, decades-old saliva would be flung out.

As for whether I'll play it when I get to Scalpsie Bay, I haven't resolved that yet. Because of the disturbance potential, if any seals are already hauled out, I certainly won't try to emulate Emily's experiments and monitor their reactions to human music. It's highly unlikely that they'd listen to my shrill rendition of 'London's Burning' with calm curiosity – far more probable that I'd flush them into the sea. When I raised the possible disruption issue with Emily, she acknowledged that there is, indeed, 'a history of people wanting to make music with animals and imposing something on them that they don't want. If you go into an aviary and make music, it can't be ignored – the birds don't have a choice. And it may come across as aggressive since birdsong's about establishing a territory. But the good thing is that if seals don't like it, they can move away.' Yet having to move away, when you're a resting and digesting seal, is, as Sue Sayer explained, categorically not a good thing as it's so energetically costly.

On the other hand, what if no seals happen to be hauled out at Scalpsie Bay? Might it be possible for me to play the recorder or even warble a tune to see if any that may be swimming in the vicinity are moved to investigate? There's a long tradition of songs for luring seals to shore, some of the melodies of which were based on, or mimicked, their vocalisations. One of the key early collectors of Scottish folk music, Patrick McDonald, includes 'A Fisherman's Song for Attracting Seals' in *A Collection of Highland Vocal Airs* of 1784, for example, and it exists as a jaunty jig on *Seal Song*, the 1981 album by the Scottish folk revival band, Ossian. Allegedly, those fishermen who were especially intent on attracting seals in order to kill them would also have a back-up plan – a basket of herring hung over the side of their boats in case the song didn't work.

Having reached Scalpsie Bay, I access the beach via a path fringed by fields of lambs bleating needily as seal pups. An information board tells me that the posts sunk at intervals into the sand are the timber remains of Second World War defences as the area was considered to be a likely landing site for a German invasion. Today's German invasion is entirely benign, consisting of a young couple holidaying from Hamburg. They're currently lugging a holdall towards the water, each gripping one of the handles, from which they extract a deflated purple dinghy. Human presence is a sure impediment to ropey recorder-playing so while they pump their dinghy up, supervised by a pair of shelducks, I put as much beach between us as possible. Luckily, the bay is wide and I aim for a distant collection of rocks of the kind that might be favoured by seals seeking a haul-out.

While the north half of Bute nestles close to the mainland, almost slotting like a jigsaw piece into the Cowal Peninsula, this south-western end juts into the Firth of Clyde and offers expansive views of another, larger island. A line of mountains grazes the sky like a serrated knife edge, the peaks flecked with residual snow. I'm delighted to recognise that this is Arran, the island on which I had my very first sighting of grey seals post-Edinburgh Fringe so many years ago.

And now, here on Bute, I'm lucky enough to be having my very first sighting of common seals, twelve on the rocks and five swimming. Seeing them is somehow both familiar and disorientating, the same but different, like meeting a friend after a decade-long gap. They're smaller than greys and seem a little more terrestrially agile, hoisting themselves from water onto rock with comparative ease. All that I've read about common seals having more delicate, cat-like faces makes visual sense now too. They also seem to be adopting a different haul-out posture – while greys often drape themselves over rocks, raising their rear flippers only when the tide impinges or they're reaching for a scratch, most of these commons are curved like croissants, heads and rear flippers held high.

For all the contrast in appearance of the two seal species, however, one thing they share is the lie that's perpetuated in their respective names. The grey's coat can be contrasting shades of black and silver, brown and cream while thanks, in part, to the common seal's vulnerability to the phocine distemper virus, its numbers have plummeted in recent decades.

The first epidemic struck in 1988, causing mass mortality around the North Sea and Baltic coasts, after which it spread around Scotland and into the Irish Sea. Related to canine distemper, manifestations of the virus include exhaustion, crusty conjunctivitis, pneumonia and an increased susceptibility to additional infections, a catastrophic development in creatures who are already immuno-compromised as a result of the build-up of industrial contaminants in their blubber. Few deaths of grey seals have been attributed to PDV and it's believed that they primarily act as asymptomatic carriers, having had contact with infected harp and hooded seals in the Barents and North Seas. Since greys have a large foraging range and use multiple haul-outs, including sites shared with commons who don't tend to be such long-distance travellers, it's thought they play a pivotal role in the transmission of PDV between common seal colonies. As a result of the first epidemic, more than twenty thousand common seals are said to have died in European waters, including hundreds right here in the Clyde and Kintyre region, while in 2002, closer to thirty thousand deaths ensued from a devastating second outbreak.

My abiding memory from the time of the first epidemic is the media attention lavished on a young musician, Fiona Middleton, based about forty miles west of here on the Inner Hebridean island of Islay, who rescued, and welcomed into her home, a succession of sick and orphaned common seal pups. As well as administering medication and injudicious cuddles, she provided entertainment, playing the violin to them and shifting her TV from wall to living-room floor so they could watch Disney videos like *The Little Mermaid*. Magazines and tabloids featured dreamy photos of Fiona fiddling waist-deep in water and there were visits from TV-am's Lorraine Kelly and hymn-singing with Harry Secombe on *Highway* too.

The German couple's purple dinghy is afloat and they've started to paddle across the bay in the direction of the seals' rocks. A black-bibbed pied wagtail bustles along the tideline, picking through the bladderwrack, and from the woodland behind the beach, I hear a cuckoo, my first of the year. Its two-note call with an interval of a descending minor third has inspired, and been imitated by, many composers over the centuries, from Handel to Beethoven to Delius. I'd probably be able to replicate it on the recorder, if only

it felt appropriate to start to play. The seals, some of them pregnant females due to pup within the next couple of months as their
breeding season, unlike that of greys, occurs during the summer,
are relaxed at the moment but a crude musical intervention would
surely change that.

Regrettably, the only published account I've come across that
offers an insight into how common seals might respond to music
consistently stretches the bounds of credibility. *Seal Morning* by
Rowena Farre, published in 1957, claims to be an autobiographical
account of the childhood years she spends in an isolated Highland
croft with her aunt and an unorthodox collection of pets including
Lora, an orphaned common seal pup. Lora seems to adapt to croft
life remarkably easily, dining on milk-soaked dog biscuits, supple-
mented by carrots and porridge. After swimming in the adjacent
loch, she's taught to fetch her mackintosh from a shelf and sit on
it in order to dry off. She also apparently learns to unpack the
picnic basket, 'first unrolling and spreading out the cloth, then
dropping a plastic cup by each of us'.

The book's unlikeliest episodes, though, concern Lora's alleged
interest in, and aptitude for, music. Farre's description of Lora's
initial reaction to her aunt's piano playing – her whole body
swaying along – seems just about plausible. The fact that, within
a week, the seal moves from tuneless wailing to being 'able to
get through "Baa-baa Black Sheep" and "Danny Boy" without a
break' and later adds 'A Nightingale Sang in Berkeley Square' to
her vocal repertoire, however, is rather more ludicrous. We're then
told she learns the blow-suck method on the mouth organ, how
to blast notes from a toy trumpet and, having been taught to hold
the beater of a xylophone in her teeth and strike the bars at which
Farre is pointing, is soon 'going over the National Anthem for the
umpteenth time'.

The climax of these musical antics comes when Farre and Lora
travel to Uncle Andrew's home near Aberdeen to participate in
one of his soirées. Naturally, Lora steals the show, outperforming
both a singer and a melodeon player. She's lifted on top of the

piano so the audience can better see her and though Farre 'stood by her side . . . in case she should be overcome by a sudden fit of nerves . . . she took the beater from me and started off with aplomb'.

In the afterword to a more recent edition of *Seal Morning*, writer Maurice Fleming asserts that though 'it took the reading public by storm' when it was first published and reviewers unanimously enthused about its evocation of a Highland idyll, a huge question mark hangs over its veracity. He tracks down a woman with whom Farre worked at a radar station during the Second World War, who casts doubts on her claims to a Highland childhood. Analysis of Farre's second and third books exposes inconsistencies too, which is bound to trigger concerns about the accuracy of her first. It seems she managed to avoid close scrutiny during her lifetime as she was uncontactable for long chunks of it, travelling, so she said, with a group of gypsies.

While I've been lingering with Farre's imaginings, the German purple paddlers have drawn closer to the rocks and the seals have clocked them. They're craning their necks to assess the level of threat and exchanging anxious glances, a hyper-alertness that chimes with my own, primed as I constantly am to respond to Dad-related predicaments. Eight would-be picnickers are making their way across the beach now too – laden with carrier bags, cool box and a windbreak, the adults are stumbling and grumbling, while the kids run, rugby-tackle one another and scream. This hugely increases the seals' twitchiness – they now need to look landwards as well as watch the firth and each other for the cue to flee.

When the picnickers are within about thirty metres of the seals' rocks and twenty metres of me, one of the women drops her bag at her feet. 'Right. We're stopping here. I'm starving.' She rummages in the bag, pulls out a disposable barbecue and thrusts it at the man standing next to her. 'If I don't get a burger in the next five minutes, I'm going to start eating people.'

The first seal, a pregnant cow, flops from rock into water and immediately, the others follow. Aware of my own twinges of hunger, I reach into my rucksack. While my right hand seeks a sandwich, my left makes one-off contact with the recorder.

First three fingers over the first three holes. Thumb over the hole on the back.

'Seals sing Star Wars tune and Twinkle Twinkle Little Star.'[3]

When I decided, after hearing the sandbank seals on the Wirral, that I wanted to learn more about seal vocalisations, this wasn't quite the revelation that I was expecting. Professor Vincent Janik, the animal acoustic communication guru at the University of St Andrews from whom Emily gleaned seal knowledge when writing *Conversation*, is having a full-on fifteen minutes of fame. Long before I met Emily, he and fellow scientist Dr Amanda Stansbury were working on a bioacoustics research project involving three young grey seals and, on the eventual publication of the outcome of their research in the scholarly journal *Current Biology*, countless media outlets have seized on the story and broadcast it worldwide.

From the post-weaning age of just a few weeks, the seals, named Zola, Janice and Gandalf, were trained, with the incentive of herring and sprat rewards, to mimic sounds played from a computer. To begin with, Stansbury and Janik focused on tones that the seals would naturally be expected to produce in the wild, then introduced notes at higher frequencies, as well as sequences that comprised vowels from human speech which would be alien to a seal. Ultimately, Zola became adept at belting out several melodies, including a recognisable version of 'Twinkle, Twinkle, Little Star'. It's very rare for nonhuman animals to have such flexible vocal capabilities – even our fellow primates struggle with learning of this kind.

Seals have a key anatomical advantage over other creatures, it seems – they possess a larynx that's very similar to ours. It's likely, too, that their social structure and, more specifically, their ritual of convening and interacting at large breeding colonies, contributes to their impressive capacity for vocal learning. While this capacity has been suspected for some time – Hoover, a common seal at the New England Aquarium in the States in the 1970s and 80s, was documented as being able to utter phrases like 'Hey! Get outta there!' – the St Andrews academics are the first to have carried out in-depth research, and the results have 'amazed' them.

When I watch the online footage of Zola articulating patterns and frequencies of sound that I've never heard a seal utter before, I'm more uneasy than amazed. Yes, part of me is fascinated by the science and especially by learning that a seal's voice box is so structurally close to our own, but another part recoils at the sight of this biddable young creature emitting abnormal yawps next to a little indoor pool. She, Janice and Gandalf have now been released into the wild as the researchers were only permitted to keep them in captivity for a year. But even so, I wonder about the impact of this year on their ability to thrive. How, when they've relied for so long on fish rewards instead of learning to forage for themselves, have they adjusted to life at sea? Has their ability to communicate with other members of their species been compromised since they were encouraged to reproduce so many foreign sounds? And, if pups are obliged to be disrupted and confined, to what extent can our pursuit of new vocal knowledge be justified?

Perhaps pre-empting this last question, Janik is quoted as implying, in many of the news stories that trumpet his research, that the outcomes could be of benefit to humans. Because the findings help us comprehend more about the acquisition and development of vocal skills, this could, in turn, lead to new insights in the study of speech disorders. I'm still not convinced that this justifies the captivity of the seals, however – we've frequently forced nonhuman animals to fulfil our needs and any positives from a research project should serve the species being studied rather than our own.

From Emily Doolittle experimenting with ways of replicating seal sounds in her music to Zola reproducing human sounds in a university pool facility, my journey into the world of seal vocalisations has taken me in directions both absorbing and unanticipated.

The direction in which scholarly research might now head is perhaps easier to anticipate, for though seals have shown they have the faculty for producing human language, they cannot, at this stage, know what the sounds they're emitting mean.

'Whether they can make sense of it,' says Janik, 'would be the next question.'[4]

11

Selkie

Everyone with whom I spoke claimed it was the worst day of rain in living memory.

Storm clouds swooped, snatched the Black Cuillin's peaks, clawed at the gabbro, guzzled all views of the Outer Hebrides. With my onward ferry cancelled, I hunkered down on the Isle of Skye, venturing out only to eat at a café-gallery where pails were precisely placed to catch drops the size of rocks from the leaking ceiling.

Though it seemed as though the rain would never ease, I woke today to a cleansed sky and a field of Highland cows squinting stoically from beneath their drenched fringes. Both wind and waves had subsided enough for the ferry to run – hence, I'm standing now on one of its breezy decks, bound for the island of North Uist. Through glasses spattered with sea spray, I'm watching the chain of the Outer Hebrides, reinstated on the skyline, grow larger and more well defined, just like my plans for this next stage of my seal journey.

A significant proportion of the UK's grey seal population is found around Scotland's coasts and the breeding colony on the uninhabited Monach Isles a few miles west of North Uist is the second biggest in the world after Canada's Sable Island, with over nine thousand pups born each year. Though I'm visiting at the beginning of June, long before the autumn breeding season, I'm still hoping to take a trip to the little archipelago some time in the next few days to learn more about the reasons for its popularity as a pupping site.

I'm hoping to learn more about the wider region's historical and prehistorical connections with grey seals too. Their bone remains have been found in middens at Iron and Bronze Age archaeological sites in the Hebrides, as well as in Shetland and along the coast of the West Highlands, while bones from an Inner Hebridean site on Oronsay date back even further, to the Mesolithic period, around six thousand years ago. For millennia, grey seals were a vital resource, providing skin for boats, boots and shelters, meat for eating and oil for lighting lamps. However, they haven't just nourished us in a practical way – my principal reason for visiting North Uist stems from the fact that grey seals have also plunged into the waters of our psyche, flippering, for centuries, through our cultural and spiritual lives. As writer and radio producer David Thomson says of seals in *The People of the Sea*, his lyrical account of his rural journeys in search of selkie lore in the 1940s and 50s, 'Land animals may play their roles in legend but none, not even the hare, has such a dream-like effect on the human mind.'

My own fascination with selkie legends began in my teenage years when I read umpteen stories featuring the conflicted character of the selkie bride. She's most commonly portrayed as a seal who emerges from the sea and sheds her skin to dance on the shore as a human for one night each year. If her sealskin's snaffled by a man who, having spied her dancing, falls deeply yet deviously in love, she's forced to remain on land in her human form. Though they may marry and produce a clan of kids with webbed fingers and fish-odoured feet, she remains enraptured by the sea. And the instant her sealskin is finally located, often by her oldest child, she reclaims it and shapeshifts her way back into the ocean.

Only later did I learn that, in addition to these tragi-romances, other versions of selkie stories exist – origin myths, for example, and morality tales in which a man who knifes a selkie is forced to journey underwater in order to be taught a radical lesson. Having inflicted the wound, only he has the capacity to heal the injured selkie and he isn't allowed to return home until he vows never to hunt seals again.

There are stories starring male selkies, too, who are usually freer than their female equivalents, able to roam at will on land and at sea. It's conceivable that many tales of dissatisfied, married women enjoying encounters with selkie lovers – tender and kind though prone to sudden disappearances – were relayed from generation to generation, yet comparatively few exist today. Nineteenth-century folklorists who collected and wrote down the stories perhaps chose to edit and omit, in line with the Victorians' puritanical attitude to female sexuality.

A rather more dangerous selkie male with a hypnotic gaze, often wearing a long, dark coat in his human form, haunts another subcategory of these stories. The woman who's lured to his subaquatic kingdom with the promise of shipwrecked jewellery and gold never returns to the surface.

Right now, I think I'd find the appearance of another passenger on this ferry's foredeck even more startling than the manifestation of a menacing selkie. With rainwater still pooled on the seats and the wave-sprayed railings off-limits to leaning, everyone's opting to stay indoors or at the sheltered stern for views back to Skye instead. I mean to stay facing into the wind for the rest of the journey, though, letting my mind cruise through oceans of selkie lore while starting to orientate myself to my new destination. North Uist is located midway along the Outer Hebridean chain and, as far as I can tell, the east side of the island looks rocky and little populated.

It's a landscape that generated one of the most notable selkie origin tales, focusing on the founding of the MacCodrum clan. David Thomson is told a version of the story, which falls into the selkie bride category, on one of his journeys to the Hebrides in *The People of the Sea*. The selkie protagonist reveals to the man who filches her skin and becomes her North Uist husband that she's the daughter of the King of Lochlann, a Gaelic name for Scandinavia. Over the years, they have many children of seal and regal lineage, and when her hidden skin's at last unearthed and she returns to the ocean, she promises to leave them fish on the shore each day till they themselves grow up, marry and expand the emergent Clan MacCodrum. Thomson is later told that seals have thenceforth remained in an ensorcelled, suspended state – 'it is given to them that their sea-longing shall be land-longing and their land-longing shall be sea-longing' – while the presence of seal blood in a man is evident from the rock 'where he sits – no matter how warm the day, when he rises the rock will be damp and the vapour from it lifting will leave crystals of sea salt beneath the sun.'

This declaration of descent from the selkies could also have arisen to account for a hereditary anomaly that might otherwise have inspired ridicule and ostracisation. Some MacCodrums were apparently born with fingers grown together so that their hands resembled flippers. Thanks to the origin tale, however, this trait could be perceived not as a deformity but as proof of the family's coveted selkie, and royal Viking, ancestry.

Indications of familial connection between MacCodrums and seals continue through the centuries. The eighteenth-century North Uist poet, John MacCodrum, wrote words to an air that may have originated as a melody sung by fishermen seeking to kill seals. MacCodrum, drawing on a tale of a bloody seal hunt on the Monach Isles, converted the melody into a lament keened by a sole seal survivor. The nineteenth-century editor of MacCodrum's poetry notes, too, that 'a woman of the same surname, and probably lineage, as the bard, used to be seized with violent pains at the time of the annual seal hunt'; while twentieth-century Canadian MacOdrums (now the more favoured spelling) insisted, based on family anecdotes, that no one of their name would have ever killed, or eaten the meat of, a seal for to do so would be an act of cannibalism.

Contemporary folk superstar Julie Fowlis, who was raised on North Uist and has spoken of its exceptionally rich tradition as far as seal-themed stories and songs are concerned, has recorded a haunting version of the grieving seal tune. She references aspects of the Clan MacCodrum origin tale in the sleeve notes of a couple of her albums too, observing that the King of Lochlann's children were hexed by an evil stepmother, transformed into seals and able to reclaim their human form only three times a year at the full moon.

Having been mesmerised, on several occasions, by Fowlis's live performances, I'm delighted that, once docked and disembarked, I can at last start exploring the island that inspires them. Leaving the harbour settlement of Lochmaddy, I weave my way through a plaid of peat bogs and lochans, hemmed by a shoreline as intricate as the selkie lore into which, over the next four days, I'll be delving ever more deeply.

By the time I check into my B&B in the early evening, a myriad other textures and colours have been added to the pattern. The white of the west coast beaches, shells ground into sand. The chestnut tinge to the corncrake's wings in the images on the visitor centre walls of the RSPB reserve I visit. The yellow of some of the wildflowers – buttercups and bird's-foot trefoil – beginning to blaze into bloom in the machair. This fertile grassland is one of Europe's rarest habitats, formed, in part, by the calciferous shell-sand being driven inland by Atlantic winds, enriching the acidic soil.

Once I've had something to eat, I'm hoping to visit another local beach that's being used as a location in an indie selkie film that's currently in production, *Mara: the Seal Wife*. The writer-director, Uisdean Murray, grew up on North Uist and I'm looking forward to speaking with him soon about his connection with the area and his fascination for selkie tales. First, though, I need to make my usual call to Dad, and I settle on the cushioned window seat in my room for loch-and-lapwing views while I phone.

'Hi, Dad. It's me.' I wait for him to ask his customary question about how things are going today in the west of Wales – once I've reorientated him, I want to tell him about the corncrake conservation project I learnt about this afternoon as it's bound to kindle his interest. Other than his answering hello, though, he's silent.

'How are you, Dad?' The silence lengthens like the June evening light. 'Dad? You still there?'

'Glowmy,' he says.

'What?'

'Glowmy.' The first syllable rhyming with cow. 'It's glowmy.'

'What is?'

'You know.'

I've grown used to his gropings for words and occasional neologisms but this feels different. He's not at all aware that he's invented this adjective and can't understand why I don't know what he means.

'Weather.'

Ah. Got it. A fusion of gloomy and cloudy. 'So was it too "glowmy" for you to go for your walk today?'

His next attempted sentence makes no sense at all. His words are slurred and jumbled, the sounds seeming not to belong to him, like Zola's weird intoning of 'Twinkle, Twinkle, Little Star'.

'Dad – Dad – stop a minute. Are you feeling okay?'

'Scurse.'

'Does one side of your body feel weak? Or numb?'

Silence.

'Dad?'

'Plost it.'

My heart's begun to tumble and plunge like the display flight of a lapwing. This speech disruption is far more serious than his usual random dementia blips.

'Dad, I'm going to get some paramedics to call round to check if you're okay. I'll ring you back as soon as I know what's happening.'

When I phone 999, my own speech is borderline garbled. On his own. Eighty-five. Father. Six hundred miles away. On his own. Speaking weird. Stroke symptom? Did I mention he's on his own?

An ambulance should be with him within fifteen minutes. I phone Dad back to tell him, relieved to find he's still conscious but unsure if he's grasping what's going on. Even though my mobile number's taped to the wall next to his landline, I'm seized by the urge to dictate it to him again, just to feel that, if he needs to, he'll be able to get in touch.

'Oh, seven, nine. . .' I begin.

'Nine, oh, seven. . .'

What am I thinking? Stupid idea. Though his words sound less slurred, he can't repeat what I'm telling him in the right order, so how on earth can I imagine he'll be able to handle the writing down?

I stay on the line till the ambulance arrives, by which time Dad seems to be spouting snatches of more sensible speech, laced with irritation at being deemed to need medical attention. Before ringing off, I give the paramedics a flustered overview of what's happened and they promise to phone me back once Dad's been fully assessed.

I draw my knees up to my chest and rest my chin on them, watching the hill beyond the loch emerge from and vanish into cloud like a mind sliding in and out of lucidity. How long will it take them to check his vital signs? To do an ECG maybe, and a neurological assessment? Ten minutes? Fifteen? Wish I could burn off some nervous energy. Wish my room had space in which to pace. Instead, I shift to the bed, do leg lifts and supine spinal twists at triple speed.

At twenty-two minutes, I'm antsier than ever. But tell myself they're just being thorough – it's good they're taking all this time.

At twenty-nine minutes, I give up trying to convince myself not to call back and badger them, grab my phone and ring.

Answerphone.

'Dad, are you there?'

No response.

'Is anyone still there?'

Nothing but a muffled *peewit* through the shut window.

I phone again.

And again.

And again.

Hospital. They must have decided to admit him. But which one? There are two possibles in the area.

Why didn't they call to let me know? No time? Too critical?

From my perch back on the window seat, the loch view warps and blurs with my tears. I feel skinless as a selkie. So helpless from six hundred-plus miles away.

Those miles will soon reduce though. I blow my nose, wipe the teary smears from my glasses and start applying myself to finding the quickest way back to Wales.

My mind spends much of the epic journey from North Uist conjuring up grim alternative versions of my first in-hospital glimpse of Dad.

Before I left, I managed to track him down in the second A&E department I contacted, by which time I was told that his speech had reverted to normal, he'd had a series of tests and would be transferred to a ward as soon as a bed became available. More comprehensive tests, including a CT scan of the brain, would be carried out in due course. The nurse I spoke to refused to discuss a diagnosis over the phone but assured me he was stable and comfortable.

Neither of these adjectives describes my own state of being when I finally crawl into the hospital, lugging a holdall packed with pyjamas, dressing gown, towels and toiletries that I collected from Dad's house on the way. A befuddling succession of lifts and corridors takes me to the ward and, at the entrance, I rub sanitiser into my hands with slow, deliberate strokes as I try to prepare myself for the shock of what I might find. The grisly gallery of images that my mind's spent hours curating, though, bears no relation whatsoever to the version of Dad that greets me – fully dressed in his favourite jeans and red polo shirt rather than hospital-gowned, and not only out of bed but striding up and down.

'What've you brought all that for?' he says, frowning at the holdall as I deposit it in front of his bedside cabinet. 'I'm not stopping here. I'm off home.'

'What?'

'Nothing wrong with me, they said. Had a bit of a turn but I'm fine now.'

'Who told you that?'

'The doctor. The one who came round – when was it? – before lunch. If you hadn't turned up, I was going to get the bus.'

I sink onto the bed, wishing I could swing my legs up onto it too and curl up under the thin, green blanket. Dad stays standing, hands on his hips. He does, indeed, look as physically well as usual, just a bit flushed with animation and hospital heat. His insistence that he goes home, though – is it based on genuine medical opinion or dementia delusion? What was the cause of his loss of sensible speech and what have the tests shown? And can it really be okay for him to be at home on his own from now on? What if the episode happens again?

'You ready to go, then?'

'Dad, we can't go anywhere till I've spoken to a doctor.'

For several hours, there's no doctor available. I feel more familiar with the routines of seal hospitals than the human equivalent just now but manage to keep both Dad's and my impatience at bay by initiating games of I-spy, hangman, and noughts and crosses. I fill him in on North Uist corncrake conservation, even though it feels like I learnt about it myself a whole lifetime ago. For five o'clock supper, I watch him polish off a plate of corned beef hash and chips, and a bowl of red jelly.

I eventually manage to intercept a doctor just as he's coming to the end of his shift. He wearily draws the privacy curtains, then nudges his glasses aside and rubs his eyes as he familiarises himself with the notes in Dad's ring binder.

'I'm Dr Shastri. How are you feeling now' – quick glance at the name board above the bed – 'Jeff?'

'It's spelt wrong.'

'I'm sorry?'

'I'm Geoff with a G.'

'Ah, I see. Well, I'll make sure we get that changed.'

'Not to worry – I'm off home now anyway.'

'That's what I hoped to talk to you about, Dr Shastri,' I quickly interject. 'Dad's discharge. And everything that's been going on since he was admitted.'

At long last, I get the details I've been craving. It's believed that Dad experienced a TIA – a transient ischaemic attack or mini-stroke. Associated with his vascular dementia condition, it's caused by a temporary interruption to the blood flow to the brain. While it doesn't cause permanent damage, there's an increased risk that he'll suffer a major stroke in the near future – to reduce that likelihood, he's been prescribed blood-thinning medication. He's not allowed to drive for six weeks and it's been recommended, too, that he have an at-home Needs Assessment with an occupational therapist. Based on her advice and with tweaks to his current domestic set-up, the hope is that he'll be able to continue living independently for as long as possible. As for his slight heart murmur, it's stable and there's no cause for concern.

The OT appointment is booked for tomorrow and, while I'm giddy with relief that there's going to be some support available, Dad's switched to full-on grumbling.

'Ridiculous. Don't need help. Nothing wrong with me.'

Mercifully, though, he consents to staying in hospital for one more night.

Over the next couple of weeks, Dad gradually but grudgingly consents to some of the recommended changes to his home and daily routine too. A lifeline alarm is installed, consisting of a base unit and wristband that can be pressed if help is needed. A daily half-hour care visit's organised and though he insists he still needs no assistance with cooking, cleaning or laundry, I can at least feel reassured that someone is checking he's remembering to take his medication. And while he balks at having a walk-in shower fitted, he agrees to an electric bath seat, which will enable him to bathe more safely.

In-between all the appointments, assessments and installations, GP surgery and pharmacy visits, my online creative writing teaching and mentoring, plus Hooper walks, I try to keep my seal project in view. While there's no immediate prospect of resuming my travels and I don't have the brainspace for engaging with research papers and reports again yet, I can still dip effortlessly in and out of the selkie theme. And even though I can't visit the North Uist locations where the soon-to-be-released *Mara: the Seal Wife* was filmed, there's what amounts to a whole selkie film fest available to watch online.

Many of these films are romantic dramas featuring a homespun cast of characters with dodgy Scots or Oirish accents and a central female selkie who, in her human form, is always young and voluptuous with pre-Raphaelite hair. The action unfolds against a backdrop of churning waves, swirling mist and forbidding cliffs and, after watching several of these films on consecutive days, I find it hard to differentiate between them.

The sun, surf and sand of *Selkie* from the turn of the millennium sees the myth transplanted to Australia and this makes for a pleasing contrast. Rather than foregrounding a young woman's sea-yearning, it focuses on an adolescent male named Jamie who initially rejects but ultimately embraces his emerging seal-self. Though it doesn't cast off the skin of cliché, it's of the Aussie teen rather than Celtic selkie type. The film successfully blends tension and humour too, refusing to take itself as seriously as some of the whimsy-laden romances. I test its family movie credentials by watching it with Dad and though the shapeshifting fantasy elements inspire some almighty sighs, the more naturalistic coming-of-age stuff has him hooked.

He's less enamoured of the 2014 Irish animation *Song of the Sea,* a poignant exploration of loss and grief from the perspective of two children, a ten-year-old boy and his mute, six-year-old sister, who, since the latter's birth, have been left on land by Bronagh, their selkie mother. While there's a beautiful simplicity to the flat, hand-drawn visuals, some of the symbolism leaves Dad baffled. 'I can't understand what they're getting at,' he says of scenes in which emotions are kept in jars with the lids securely fixed, and switches his attention to playing tug-of-war with Hooper.

It's just as well that I watch *Ondine*, the 2009 movie directed by Neil Jordan, on my own, as Dad would find its shift between myth and gritty thriller both irritating and perplexing. Colin Farrell plays Syracuse, an alcoholic Irish fisherman who catches the titular heroine in his trawl net and falls in lust with her. His young daughter, Annie, wheelchair-bound and waiting for a kidney trans-plant, is convinced Ondine's a selkie and within this mythical context, a host of social and political issues from bullying to immi-gration are introduced. It's a testament to how immersed I've become in all things selkie that I find Annie's explanation of Ondine's origins far more credible than the one that's given in the film's violent denouement – it turns out that she's a Romanian drug mule with a rucksack packed with heroin, on the run both from the law and her brutish pimp.

As I watch these films and mull over the varied metaphorical uses to which contemporary filmmakers have put the selkie myth, I consider what function the tales might have served in Hebridean and Orcadian societies in centuries past. It seems possible that, as with *Song of the Sea*, they could have acted as a means of exploring and palliating feelings of grief. Death in childbirth would have been an all-too-common occurrence and the trope of the selkie wife returning, like Bronagh, to the ocean could have provided a palatable explanation for the tragedy of a missing mother. The threat of death by drowning would have also been a constant, and the belief that lost fishermen might transform into selkies could have lessened the fear felt by family members.

The figure of the male selkie might have had a role to play on the occasion of a woman becoming pregnant by a man other than her husband too. Passing it off as the result of a seduction by a virile visitant from the sea could have protected reputations and helped maintain the peace in a small, close-knit community.

As for other purposes that the myth latterly serves, Jungian psychoanalyst Clarissa Pinkola Estés tells, and gives a beguiling interpretation of, an Inuit selkie wife story in her bestselling bible of female empowerment, *Women Who Run With the Wolves*. It's a book that offers, through retellings of myths and fairytales from a range of different cultures, a road map by which women might navigate away from the stricture of socialisation, and travel back to their wild, intuitive, creative selves.

I was first introduced to Pinkola Estés's book on the day I left the UK to take up my writing opportunity in Australia. It was a late afternoon in late June and Dad had generously driven me and my bulging backpack from Wales to Heathrow. Wimbledon radio commentary had filled in the gaps in our conversation, mine a stuttering mix of excitement, hope that heading overseas was the right option and sadness at the impending goodbye. Several hours earlier, I'd bid farewell to Mum. She'd chosen to stay at home rather than travel to the airport, insisting she needed to look after their elderly terrier, though I suspected it was mostly disapproval at my latest career move that had influenced her decision not to come.

In Terminal 4, Dad and I were joined by two of my closest friends. After I'd checked in, we all sat together in a café where Hannah handed me a travel journal plus a pen patterned with a map of the world and Angela gave me a copy of *Women Who Run With the Wolves*. On the title page beneath the subtitle, *Contacting the Power of the Wild Woman*, she'd written 'For Sue, one of the most truly wild women I know.' Touched though I was by both gift and message, the wildest thing I could currently think of doing was working out how to cram such a hefty book into my overloaded hand luggage.

It was too big to fit in the pocket of the seat in front of me too so I spent most of the twenty-six-hour journey with it open on my lap. At some point between my stopover in Bangkok and change of plane in Melbourne, I came upon the selkie story in the 'Homing' chapter and proceeded to read it at least five times. There are many ways in which a woman can lose her sealskin,

Pinkola Estés suggests, including by pursuing what others think she should do instead of what she knows she must. Retrieval of the skin symbolises a return home – not to an actual location but to 'the pristine instinctual life that works as easily as a joint sliding upon its greased bearing.' By the time I touched down in Adelaide, I was convinced that I'd left my physical homeland with the subconscious goal of learning to fully inhabit my authentic 'seal-skin/soulskin'.

As I start to do some research into other ways in which, and other people for whom, the selkie myth seems to resonate, Dad continues to adjust to his new routine. The lifeline alarm takes some getting used to – he keeps removing his wristband at night and depositing it on his bedside table in such a way that the sensor thinks he's had a fall, which triggers an emergency response call via the base unit. The daily care visit seems to be working well, though, even if Dad's treating it mostly as a social call, while I'm just relieved there's someone else apart from me who's on alert for signs of a stroke.

I'm also relieved, not to say surprised, to find that, in spite of assuming I wouldn't easily meet people who are engaging with the selkie theme now that I'm back in Wales, I come across Rachel Taylor-Beales, a folky-jazzy-bluesy singer-songwriter and theatre-maker. Her work happens to be steeped in selkieness and, as she lives within an hour of Dad, I take up her invitation to call round so we can chat about it.

'Are you okay with dogs?'

When I arrive, there's a bout of barking followed by panty excitement from Megan the chocolate Labrador, a shorter and marginally rounder version of Hooper. I bend to take the shoe she chooses to bring me from the pile by the door and scratch her under her chin. 'Mine does exactly the same as this whenever there's a visitor.'

The three of us move to the kitchen and while Megan's given a yoghurt pot to lick out, Rachel makes tea and lifts a giant jar of biscuits down from a shelf. Her accent is an interesting hybrid of standard English enhanced with slightly more expansive Australian vowels.

'I grew up between Fremantle in Western Australia and Nottingham,' she explains, adding that the contrast between living by the ocean and in the landlocked East Midlands gave her an early sense of selkie-type dislocation.

Megan emerges from the yoghurt pot with most of what remained smeared on her nose, and her tongue unable to reach it, another long-standing Hooper trait. She follows us past Rachel's home recording studio, with guitars of contrasting sizes hanging on the walls, and into the lounge. Already feeling like we've bonded over our dogs, I'm really looking forward to hearing about Rachel's acclaimed album, *Stone's Throw – Lament of the Selkie*, and the multi-media show of the same name for the development of which she recently won Arts Council of Wales funding.

Tucking an amber strand of hair behind one ear, she takes in a breath and begins to recount the concept behind both show and album – how she's fused myth with autobiography and used the archetype of the selkie wife to explore a series of seismic personal challenges. While doing a sound check for a gig in Italy in 2012, she fell from the stage onto marble flooring: she was twenty-four weeks pregnant at the time and though her baby was unharmed, she herself suffered serious back and hip damage. During the long period of near-immobility that followed, her muscles became so atrophied that she was unable to sing, while back spasms prevented her from holding a guitar for more than half a minute at a time. After a harrowing birthing experience, compounded by her injuries, she also endured episodes of memory loss. 'We still don't know what it was – perhaps a mini-stroke or maybe a reaction to trauma. One episode was very scary. I was lucid enough to know where I was but I couldn't recall what had been happening and that I'd already given birth to my daughter.'

When she finally started work on a new album, she admits that she was 'still in a place of crisis. I needed an anchoring theme and I chose the selkie as a safe, creative place for exploration.' The selkie's entrapment on land in human form reflected Rachel's own sense of being trapped by her injured body. 'And it also let me be a bit ambiguous.'

'In what way exactly?'

'Well, for one thing, there's ambiguity when she re-enters the ocean. It could be very positive, a returning home. On the other hand, if it's a woman with mental health issues or post-natal depression, it could also be seen as suicide.' Rachel pauses to gather both her thoughts and a ginger nut from the jar on the coffee table in front of us. 'I wanted to explore the version of myself that didn't recover. My own "what if". What if I hadn't got better? What if I'd stepped into psychosis?'

She lets the questions bob in the space between us like the heads of surfacing seals. It's an affecting moment, interrupted only by Megan's barking at the arrival of mail on the hall floor.

This prompts Rachel to go on to describe the arrival of another vital theme in her creative letterbox. 'The refugee crisis in Gaza was huge at the time that I was making the album and it affected me deeply,' she says. In the title song, 'Stone's Throw', she aligns the loss and sea-longing experienced by the selkie who's obliged to live for many years on land with the grief of displaced people and the emotional chaos caused by enforced migration.

As well as interweaving this socio-political strand with the mythical and autobiographical, she aims to integrate other women's voices and birthing stories into the *Stone's Throw* live show on which she's working. 'After I wrote a series of blog posts about what I'd been through, I began to receive messages from women all round the world who connected with what I was saying and needed to share their experiences. These stories are often taboo and stay untold so I wanted to make a space where they could be heard.'

Balancing all of these themes feels like an ambitious task and I'm pleased that, soon after we meet, I have the opportunity to catch a performance, the culmination of the funded research and development phase of the project. Perhaps in recognition of the harrowing subject matter, audience members are comforted in advance by custard creams in the foyer, then seated in the theatre at bistro-style tables with candles. Alongside her fellow actor and against a backdrop of marine-themed visual projections and verbatim recordings of women's thoughts on when and how their skin best fits, Rachel sings and stories her way through the show with warmth and courage. She showcases her skill as a multi-instrumentalist too, playing keyboard, ukulele, acoustic and electric guitar.

Judging by the post-performance discussion, many of the audience members are stirred by what they've seen and at least one woman is plainly shaken, proposing that the show needs to carry a trigger warning. The selkie acts as an intermediary between the audience and Rachel's pain, however, making the witnessing of it more endurable, and the emphasis on post-trauma regeneration is uplifting too.

As Rachel stressed when we met, 'The selkie's all about cycles' – seasons on land and seasons at sea. Telling herself 'this is just for a season' was very consoling when her mood-tide fell perilously low.

Travelling an hour to spend an afternoon chatting with Rachel about selkies is one thing. Heading back up to the far north of Scotland to resume my seal journey is another. I'm of course itching to be there but at the same time, I'm hyper-anxious about leaving Dad. Whenever I've spoken about it to Bethan, the carer who calls on him most regularly, she's insistent that I should go – 'Your dad's doing well at the moment, babe. We know how to get hold of you. You'll hear from us if anything's wrong.' Dad, himself, is happy for me to leave again too – 'For god's sake, go wherever it is you want to go. You don't need to worry about me.' But still, I'm dithering.

This afternoon, I've brought him to the local park for a walk and we're following a paved path round the lake. It feels very tame compared with the hill and coastal hikes we used to do but considering that he's in his mid-eighties and a month ago, as I hared to the hospital, I was having visions of him being forever bed-bound, it's a resounding triumph. Hooper's with us too, straining on the lead, convinced that the food pellets we're scattering on the water for the mallards and Canada geese are actually meant for him.

'Look! There's a wotsit!'

I follow the direction of Dad's finger.

'There! In the reeds – you know – the one with the red bill.'

'Moorhen.'

'That's it – moorhen.' He smacks his forehead. 'This bloody memory.'

In fact, there are two adult moorhens beaking aquatic insects, and three downy black chicks cheeping in their wake. As we watch them, I go through a mental checklist of the extra measures I've put in place to try and pre-empt a Dad-related crisis if I travel north again. I've restocked the freezer and mapped out a weekly menu. I've stuck a notice on the inside of the front door to stop him going out for a walk at the time the carer's due. I've triple-checked the date of his follow-up appointment at the stroke clinic. I've made sure his surgery's registered the consent form I filled in that'll allow me to discuss his health and any test results with his GP. Can't think of every eventuality though. Leaving will still be a risk.

A swan clambers out onto the bank, seeking the source of the pellets, leaving his mate and six cygnets lunging for the food we've already thrown. When he sees Hooper, he lets out a hiss.

'Show-off!' says Dad.

Pleasant though it is to amble round the lake, it's reminding me how much I'm craving waves instead of pavement, bickering seals instead of crotchety swans. So why can't I just go back to Scotland, knowing I've done as much as possible to help Dad for now? Why can't I steel myself to deal with the only-child guilt and anxiety that leaving will bring? I think of Rachel Taylor-Beales rebuilding and renewing, refusing to allow the trauma she'd experienced to hold her back.

Right now, neither Dad nor Hooper can be held back from the ice cream kiosk that's adjacent to the lake. Twelve flavours are listed on a sign in the window which induces, from Dad, a worried frown.

'I'll just have an ordinary one, thanks.'

We carry our vanilla cones to a low wall and perch there.

'So when are you off then?' he asks.

'Off where?'

'Wherever you're going.'

'Scotland. And I'm not sure if I am.'

There's a long silence as Dad focuses on licking all round the circumference of his cone so the scoop of ice cream stays smooth and even. My mind spends the silence rewinding to that other June leaving, the Heathrow afternoon when I was Australia bound. Though I was a little conflicted about going, it was so much easier to do so in my unencumbered twenties and, once guided by Pinkola Estés, I never looked back. This late June leaving, if that's what it turns out to be, is much less straightforward. Until the last few weeks, I've never given much attention to the choice the selkie must make between staying with her children on land and returning to her soul-home of the sea, yet I now think of it as deeply complex and painful.

'Don't think I've ever been to Scotland,' says Dad, offering his fingers to Hooper for a lick of vanilla drips. 'We'll talk from time to time, won't we, when you're there?'

'Of course. I'll phone twice a day, Dad. I always do.'

'Do you?' Eyes wide with surprise, he pops the end of his cone into his mouth, then smiles. 'Well then, that'll be lovely.'

12

Skins of Silver, Chains of Gold

I'm on an island off an island known as Mainland, wearing the fleece, gloves and hat that would usually be mothballed in summer. While I'm pleased to have resumed my seal journey and got myself this far north, I'm also feeling a sense of dislocation, much of my headspace still taken up with Dad-related concern instead of immersion in my new locale.

On the map, the Orkney archipelago looks like the pattern of blotches on the side of a cow seal's neck by which she can be identified. Having made it to the northwest edge of the biggest of these blotches, I've walked across a tidal causeway to the Brough of Birsay, a dot of an isle, the small size of which belies its historical and ecological significance. I'm currently wandering among a network of low stone walls topped with grass and thrift – Norse ruins from the ninth century onwards that include several long-houses and a smithy. There's apparently a well, signifying an earlier Pictish settlement, to be found here too.

I clamber higher to a squat, white lighthouse with castle-like crenellations and take in the view from the cliffs on the seaward side of the island. Puffins are flurrying between sea and land, bills crammed with sandeels, stubby wings beating to a blur. After disappearing into their burrows to feed their chicks, they re-emerge for a brief preen, then fling themselves towards the sea again, the orange paddles of their feet splayed out behind them. These feet and their flamboyant beaks bring radiant relief to the scene's many shades and shapes of grey: careering clouds, simmering sea, outlines of islands lying to the north – plus the heads of sporadically surfacing seals.

More than a third of the UK's grey seal births are believed to occur in Orkney, with various of its seventy islands playing host to pupping nurseries each autumn. Yet, as I'm already discovering, there should still be plenty of opportunities to spot seals during this summer visit as they frequent the islands' waters all year round. I'll hopefully be able to delve more deeply into selkie mythology while I'm here too. Like the Outer Hebrides, Orkney is a key location as far as the origin and evolution of these legends is concerned and, rather than refer to a guidebook or travel website, I've decided I want traditional selkie tales and ballads, and contemporary selkie novels, to guide me through the Orcadian landscape.

My first stop, shortly after I'd wobbled off the ferry that jounced across the Pentland Firth from its terminal close to John O'Groats, was the immense Neolithic stone circle of the Ring of Brodgar. On an isthmus between lochs, twenty-seven of the original sixty stones stand resolute against the wind while wisps of bog cotton quiver at their base. I explored the site with a pivotal episode from Joan Lennon's 2015 genre-defying young adult novel, *Silver Skin*, running through my mind. With its elements of science fiction, fantasy, folklore and historical romance, it features Rab, an adolescent boy from the far future whose silver suit with a computer downloaded into an arm panel facilitates time travel back to Stone Age Orkney. Though actual seals make only the briefest of appearances, Lennon's fresh take on selkie legends is integral to the development of the narrative. Cait, the young woman who witnesses Rab's bumpy arrival and subsequent expressions of pain (*'It must be the seal language'*) is convinced he's a selkie. Meanwhile, the community's elderly medicine woman, Voy, steals and hides his silver suit in the hope that he, and it, will bring luck and other benefits, including a softening of the increasingly savage climate.

The Ring of Brodgar episode is the most visually arresting of the novel with a ceremonial gathering of clans from across the region and a procession to the centre of the circle in order that offerings can be made. Voy's contribution is Rab's silver skin and, when he sneaks back later to try to retrieve it, there's a fierce communal confrontation. As I paused in a gap in the circle where a stone would once have stood, my eyes conjured Lennon's lightning strike that splits one of the megaliths, then deflects onto Rab's skin while Voy holds it aloft, jeopardising his chance of getting back to the future.

Compared with compromised time travel, my journey from the Ring of Brodgar to the Brough of Birsay couldn't have been more straightforward, taking less than half an hour. Now, from my wind-walloped spot on the island's west-facing cliffs, I watch the waves hurtling in, hurdling over each other in their eagerness to reach the shore, and flobs of spume being flung into the air to accumulate on the cliff top. A bull seal surfaces and submerges several times over, perfectly at ease in the surge and heave of the sea while a juvenile bobs up closer to the cliffs, craning her neck to check out the bull until a wave breaks over her quite spectacularly.

A succession of visitors clambers up the slope from the Norse ruins, joining me at the island's peak. First, a woman with a child at the end of each arm, the younger girl pulling ahead, the older boy lagging behind. Then, a man in a brown duffle coat who fiddles with the bottom toggle as he regales me with tales of wild swimming. A rather harassed-looking guy pants his way to the top too, a woman following at a sullen distance. He casts his eyes urgently over the cliff, then lets out a breath of relief. 'Thank Christ for that. I told her there'd be puffins.'

With the wind gusting so forcefully, no one's inclined to stay for very long and I have ample solo moments in which to watch the seals and also contemplate what lies beyond the western horizon. Some forty miles away is the tiny island of Sule Skerry, a storm petrel and gannet hang-out, once home to Britain's most remote manned lighthouse but now best known for its appearance in a mournful ballad. Versions of 'The Grey [or Great] Selkie [or

Skins of Silver, Chains of Gold

Silkie] of Sule Skerry' from Orkney and Shetland have been recorded by a whole range of folk artists from Joan Baez and Judy Collins to June Tabor and Maddy Prior. The most intricate variation of the narrative concerns a young maid who's unsure as to the whereabouts of the father of her baby. One night, he reappears, outs himself as a selkie and suggests she should look after their child for a further seven years. At the end of the designated time period, he returns to claim his son, bringing money for the maid plus a chain of gold for the selkie boy's neck. As a parting shot, he makes a chilling prophecy – that the maid will marry a 'gunner good', a hunter who will kill both selkie father and son. This does indeed come to pass, as confirmed by the gold chain with which the distraught mother is presented by her new husband.

The typical duality of the selkie – 'I am a man upon the land; / I am a selkie on the sea' – is enhanced by the fact that he dwells on and around the liminal, sea-washed rock of Sule Skerry. Listening to the ballad always leaves me stranded in a liminal emotional space too, drawn to the beauty of both of the melodies to which the lyrics have been set but repelled by the sense of relentless tragedy.

A more benign selkie narrative with Sule Skerry as the setting is found in storyteller Tom Muir's collection, *The Mermaid Bride and Other Orkney Folk Tales*. Two fishermen from Sandwick, the parish through which I travelled to get here from the Ring of Brodgar, are blown off course, ending up at Sule Skerry. There, in a little house, they come upon a woman who had disappeared from their community some time ago. Shortly after their arrival, a seal enters the house and heads past them into a separate room and, a quarter of an hour later, 'a fine big man in grand clothes' emerges from the same room to welcome them to his home. The story tails off with an unexpected lack of tension: the woman and her selkie husband simply share a meal with the fisherfolk who return to Sandwick the next day without attempting to wrest her back there with them.

After one last scan of the murky horizon in the vain hope that I might catch a glimpse of Sule Skerry, I likewise decide to return to Sandwick, outpacing the tide as I cross the causeway, in search of some food and my B&B.

Contemporary selkie novels that are not just set in Orkney but in Cornwall, Yorkshire, Maritime Canada, the east coast of the USA. Rite-of-passage narratives that use the shapeshifting selkie to engage with the physical and emotional flux of adolescence. LGBTQ fiction in which seal–human metamorphosis enables exploration of sexual orientation and fluidity. Novels that seek to understand other physical transformations that a body may undergo, such as the ravages wrought by anorexia. Though none has addressed the bewildering transitions of dementia yet, it's remarkable how far the selkie has travelled from its geographic and allegorical roots.

I've lugged a fair few selkie novels with me on this section of my seal journey and, having eaten and checked into my B&B, I spread them on the bed across the heather-patterned duvet. Before reconnecting with any of them, though, I make my usual evening phone call to Dad, heart careening till I hear that he's okay and an emergency return to Wales won't be needed. We enjoy a lengthy chat: while he can't quite recall if he watched *Pointless* earlier, his memory readily unearths facts about Orkney from school history lessons and news reports. The scuttling of the German naval fleet in the harbour of Scapa Flow after the end of the First World War. The building of the Churchill Barriers to protect British ships anchored in the same harbour during the Second.

Phone call done, I turn my attention to my hoard of novels. On the Orkney side of the bed, alongside Joan Lennon's *Silver Skin*, is *Song of the Selkies* by New Zealand writer Cathie Dunsford which I've just finished reading. While its theme – a storytelling group of women travelling to Orkney after performing at the Edinburgh Festival – is appealing, and the celebration of sexual diversity fits seamlessly into the selkie narrative, I can't help but feel nettled by the natural history gaffes – puffins nesting and bluebells blooming in what must be, since it's post-Festival, late August – as well as the underwater seal chat in dodgy Scottish accents ('I doot I'd have believed it meeself').

By contrast, the dialogue and atmospheric landscape descriptions in Simon Sylvester's 2014 thriller, *The Visitors*, are pitch perfect. The selkie theme again facilitates an exploration both of female sexuality and the lurching journey from adolescence to adulthood, and the mix of myth and murder is skilfully managed. Set on the fictional Hebridean island of Bancree, the novel stars seventeen-year-old Flora who sets out to investigate a string of disappearances from her community. Her transformation from lonely, bullied school-girl into plucky and empowered young woman is as stirring as any subsidiary character's shift from seal into human.

In some of the selkie novels I've read, the pivotal shapeshifting episode takes place offstage, while in others, it's described as a simple slipping from a skin, a peeling and parting from the crown downwards. Only a few writers seem to focus on the process and physicality of the change, the sensation of feet webbing into flip-pers, the instant acquisition of flab and fur. One such writer is Australian Margo Lanagan: in *The Brides of Rollrock Island*, she gives lyrical voice to six different characters including the witchy Misskaella, who, for a fee, will call forth a meek and beautiful woman from inside the body of a seal to satisfy an islander's desire, and Daniel, one of the many resultant selkie sons. 'I don't know how to tell it all,' he says of the at-sea portion of his life. 'The best I can do is overlay a skin of man-words on the grunt and urge and flight and slump of seal-being.' When transferring back to land he struggles to remember how to walk: 'My feet dragged and my legs tried to rescue them – how was I to support myself and balance, on these two stalkish things?'

In her coming-of-age novel, *Deep Water*, Lu Hersey takes descrip-tions of physical transformation one stage further, with palpable parallels with the upheavals of puberty and an emphasis on the pain of the experience as well as its strangeness. At the beginning of the novel, fifteen-year-old Danni has no knowledge of her selkie heritage and, since her mother has gone missing, she has no one to guide her through the onset of her weird, briny dreams or the welling up of water from the lines in the palms of her hands. As the magic and mystery of the narrative unfolds, she learns to embrace the new abilities that selkieness confers as well as accept the bodily discomfort – 'This time I'm more aware of the pain. The lengthening and fusing of bone, the skin hissing and fizzing as it becomes a part of me.'

My eyes come to rest on the final selkie novel with which I decided to travel, *Seal Woman*, an unexpected foray into fiction by Ronald Lockley. Described by the *Times Educational Supplement* as 'a curious, passionate masterpiece' following its publication almost half a century ago, it features the enigmatic Shian, a woman who becomes increasingly seal-like as the novel progresses and to whom the narrator is irresistibly attracted. Together, they haul out with a seal colony on an offshore skerry and, for a while, adopt the seals' annual mating and moulting rhythms. Though the novel has strong fantasy credentials, it's also saturated with Lockley's knowledge of seal behaviour and biology, attributes that are almost entirely missing from every other selkie tale I've read. He brings a brutal dose of realism to the ending too with the narrator returning to the coast some time on from his affair with Shian to discover that hunters have killed all the pregnant cows and a 'seal with a human face and hair' is presumed to have been among the victims. While other writers use the selkie as a means of safely exploring adverse human experiences, Lockley seems to be aiming to school a wider audience in the persecution suffered by seals.

Next day, I reach my seal trip's Furthest North.

With the wind pausing for breath and the sky a fitful blue, I take a boat to Westray, one of the grey island shapes I spotted from the Brough of Birsay. Home to just six hundred residents, it's small in terms of surface area but its coastline squiggles and jinks for a full fifty miles. In the Heritage Centre in the main settlement of Pierowall, I learn about the kelp gathering that used to be a core local industry, and gawp at the Westray Wife, a tiny, carved Neolithic figure that was unearthed on a nearby archaeological dig and garnered world renown. I walk across machair and on shrieking, reeking seabird cliffs and, for one sun-blessed hour, I walk in just a T-shirt, with waterproof stuffed in my rucksack and fleece knotted round my waist. I beware of the bull, straddle ladder stiles, watch the spearheaded dives of gannets and picnic on bannock and crumbly Grimbister cheese.

I also consistently spot grey seals at the coastal locations I visit, though not at Grobust beach, the one place where I was told a sighting would be guaranteed. I arrive late in the day and spy just a couple of kayaks resting on the white sand, albeit with the bow of one nosing the hull of the other, in suckling-pup-and-cow formation.

It was from Westray that a number of sightings of kayak-paddling 'Finnmen' were made in the late seventeenth century, potentially leading to extra elements being incorporated into the prevailing selkie folklore. With varying degrees of plausibility, it's been claimed that these visitors to Orkney waters were either Sámi, the indigenous people of Northern Scandinavia, or Inuit from Greenland. In both cases, their kayaks, unlike the lightweight polyethylene versions here on Grobust beach, would have been made of sealskin, a fact that's recognised by John Brand, a minister who took an investigative journey to Orkney and beyond on behalf of the Church of Scotland. In *A Brief Description of Orkney, Zetland, Pightland-Firth and Caithness*, published in 1701, he writes that 'There are frequently Fin-men seen here upon the Coasts . . . and another within these few Months on Westra. . . His Boat is made of Seal skins . . . and he sitteth in the middle of his Boat, with a little Oar in his hand, Fishing with his Lines.'

While Brand stresses that if anyone ever tried to approach one of the kayakers while on the water, he would paddle away from them at speed, there have also subsequently been suggestions that Finnmen would have periodically needed to step out of both their sealskin kayaks and their waterlogged clothing to allow everything to dry and regain buoyancy. It's easy to imagine how details of strangers divesting themselves of their animal skin garments and stretching out beside them on the rocks might have lent additional colour to the selkie tales that were already in circulation.

As I jettison my walking boots and enjoy the barefoot feel of cool, soft sand, I recall a rather more curious detail that accompanies a report of a Finnman observation. In *A Description of the Isles of Orkney*, the Reverend James Wallace mentions that, following a 1684 sighting of a Finnman off Westray, the locals 'got few or no fishes; for they have this Remark here, that these Finnmen drive away the fishes from the place to which they come.' This fish-scarcity claim in the aftermath of a Finnman's appearance reminds me of the argument that the fishing industry has vehemently made, over many decades, about the perceived proliferation of seals in UK waters. Seals are believed to constitute a rival fishing fleet, which directly competes for available fish stocks and thereby reduces the potential profit of the industry. Objection to their presence peaked in the late 1970s when Orkney was embroiled in a controversy – a government-sponsored cull and a high-profile anti-slaughter campaign – that was every bit as tangled as the strands of kelp over which I'm currently stepping.

The proposed cull was by no means unprecedented. Seals had been commercially hunted for their skin and oil for many years, with the Monach Isles and Sule Skerry among the most heavily targeted locations. The adaptive attributes – waterproof fur and blubber – that enable grey seals to thrive in the marine environment, were the very features that made them so susceptible to exploitation and, by the early twentieth century, it was estimated that there were only about five hundred remaining around the whole of the UK. As a result, the Grey Seals (Protection) Act, decreeing several hunt-free months to coincide with the breeding season each year, was introduced in 1914: it was the first time any mammal had been legally safeguarded by parliament. A later version of the legislation, the Conservation of Seals Act, extending protection to the common seal for the first time, was passed in 1970.[1]

It again prohibited killing during the breeding season yet still permitted seal-shooting licences to be issued if fisheries' interests are deemed to be threatened and for scientific and population management purposes.[2] During the late 1950s and 60s, grey seal culls were periodically carried out as population control measures but it wasn't until Orkney was in the spotlight in 1978 that locals, the wider public and national conservation organisations united in a large-scale protest.

At that time, it was alleged that grey seal predation on fish was annually costing the industry up to £25 million.[3] Further economic pronouncements were made about the fact that the grey seal is the final host of the parasitic codworm, ingesting fish infested with the larvae, then excreting the eggs laid in its stomach by the mature adults. When these eggs hatch, the larvae are eaten by crustaceans that are, in turn, consumed by cod and other bottom-feeding fish. Although the codworm is harmless to humans, infested fish are regarded as unpalatable and therefore have a lower market value, sparking the scientifically dubious claim that fewer grey seals in the ocean would result in fewer worm-riddled cod being present in commercial catches.

As a result of these arguments, a government-sanctioned plan was advanced whereby a swingeing cull of adult cows, newborns and moulted pups in Orkney and on the most remote of the Hebridean islands, North Rona, would be carried out during the closed breeding season each October over a six-year period with the aim of reducing Scottish grey seal numbers from over fifty thousand to thirty-five thousand. Opponents of the cull, including various conservation bodies, advocated for more detailed scientific studies of seal diet and its impact on fish stocks. Greenpeace took a more interventionist approach, indicating that it viewed the cull as an attempt to shift blame for declining fish stocks from human overfishing to the seals and despatching the *Rainbow Warrior*, with journalists and ITN reporters on board, to shadow the boat from which the seal-killing would take place. Thousands of letters objecting to the cull were sent to Prime Minister Callaghan from concerned members of the public throughout the UK, while in Orkney itself a protest group called Selkie was formed. Its members devised a logistically complex but effective campaign, setting up camp in often grim weather conditions on the remote pupping islands with the intention of creating a disturbance when the hunters approached so that the adult seals would move into the water and avoid being shot. This mix of local and national protest finally induced the Secretary of State for Scotland to announce the suspension of the cull due to public concern without a single seal having been killed.

Among the widespread jubilation, there were, of course, some who were dismayed at the decision, with conservationist Sir John Lister Kaye the most unexpected detractor. In his 1979 volume *Seal Cull: The Grey Seal Controversy*, he argued that public opinion was misinformed, swayed by propaganda and sentimentalism ('The man in the street had no alternative but to identify the wet eyes [of the pups] with those of his dog staring balefully up at him from the carpet beside the television'). By contrast, the government and its advisers were 'acting in what they considered to be the best interest of mankind, the fish in the sea and even the seals themselves.' Because of the fact that the grey seal had 'come back with a vengeance' in recent years, he supported the concept of a more covert pelagic cull, out of the public eye – 'In many ways the concept of a British sealing vessel constantly cruising round removing seals, although costly, might be better for seals and for science.'

In spite of the success of the 1978 Greenpeace–Selkie campaign, the government's Orkney culling programme resumed the following year and continued until 1981, albeit with fewer seals being killed than was originally mooted. Since the early 80s, however, a more realistic model of interactions between seals and fisheries has been advanced, one that isn't simply fixated on the assumed link between seals' prey consumption, their population size and the magnitude of commercial catches. Since fish are also eaten by seabirds, other fish and cetaceans, all co-existing in a complex multi-species web, it's naive to believe that a seal cull would automatically lead to more fish of the appropriate age classes and species becoming available for procurement by the fishing industry.

More recent science, centred on a 2010 study in the Gulf of Maine, has demonstrated how grey seals and other air-breathing marine mammals play a vital, fertilising role that ameliorates the health of the entire ocean environment. The concept, referred to officially as 'the whale pump' and more frivolously as 'the poop loop', describes how the diving and surfacing behaviour of pinnipeds and cetaceans creates an upward movement of nutrients through the release of faecal plumes. This stimulates the growth of phytoplankton, microscopic plant life that forms the foundation of the whole marine food web. The ecosystem beyond the ocean benefits from the process too as, through photosynthesis, phytoplankton removes carbon dioxide from the atmosphere and produces oxygen in exchange. Culls that imprudently target seal populations can therefore have a far wider impact than has traditionally been imagined, causing a decline in the faecal nutrients that enable phytoplankton to flourish at the very time when its CO_2 absorption properties are needed to help counter the galloping pace of climate change.

Regardless of all these findings, while there has been no major authorised cull of grey seals anywhere in the UK since the early 1980s, deep antipathy towards the species on the part of the fishing community has persisted and I'll be visiting the site of a more recent conflict on the next stage of my journey. For now, I sink down onto the sand, rummage in my rucksack for the slice of Westraak tart I bought earlier – *aak* is the local word for guillemot but this is mercifully just fruitcake filling in a pastry case – and reflect on the time I've spent in Orkney. Having dithered for so long over whether to travel because of Dad's health, I'm grateful to have made it here and to have experienced a location that has brought into even sharper focus the duality of our attitude to grey seals. Scapegoated and vilified on the one hand, yet embedded in our imagination and cradled in myth.

13

The Scramble for Salmon

The village of Crovie quivers on the brink of the Moray Firth in northeast Scotland, its line of cottages so crammed between cliff and sea that there's no room for roads, with residents obliged to transport provisions in a wheelbarrow from the edge-of-the-settlement car park. It's still early in the day and, as yet, there's no sign of movement from either locals or any of the visitors who come to stay in the former fishing cottages, lured by the prospect of summer holidaying in such a unique location.

Along with much of the North Sea coast, Crovie suffered a devastating storm surge in 1953, prompting many of its residents to abandon their battered homes and move the short distance along the firth to its big sister fishing village of Gardenstown. I follow the direction of their flight, picking up the cliff-top path and pausing every few hundred metres in the hope of spotting a seal or the flashy leap-and-splash of a bottlenose dolphin. Over the years, Gardenstown's expanded vertically rather than horizontally and my visit begins among twenty-first-century new-builds, a petrol station and Spar before steeply dropping several centuries down through a huddled muddle of houses, churches and narrow stepped streets as far as the harbour. I'm reminded of a seabird cliff with each breeding species occupying a different level: the variously aged buildings cling to the terraced ledges much like nesting guillemots and kittiwakes.

At the harbour, after sidestepping a weave of ropes, lobster pots and buoys, I pause at a wall daubed with *God Is Love* graffiti. Its claim and the harbour's current calm contrast with the impression I gained of both Gardenstown and Crovie from my pre-trip research into a series of impassioned confrontations over the shooting of seals. In 2012, Usan Salmon Fisheries Ltd, a family-run business describing itself as one of the last wild salmon netting companies in Scotland, received unwelcome press coverage when an English couple cut short their holiday in Crovie, in distress at the shooting of fourteen seals, allegedly by Usan, with a number of the carcasses left on the shoreline.[1] In the previous year, the Scottish government had introduced a seal licensing scheme, requiring any fisherman who wanted to kill seals that were raiding his nets and protect what he considered to be his harvest to apply for a licence. The application would be assessed based on scientific advice, and a maximum number of seals that could be shot in that particular location as a last resort if non-lethal measures proved ineffective was stipulated. In response to the outcry in Crovie, Usan's director, George Pullar, contended that the company 'operate[s] according to strict licensing conditions and protocols imposed by Marine Scotland', and in 2012, though the couple's claim was investigated by the police, Usan were deemed not to have contravened any of these regulations.[2]

In the past couple of days, whenever I've mentioned the licensing scheme to random folk on my journey south from Orkney via Inverness and my subsequent arrival here at the Moray Firth, I've never failed to get a reaction. Some have been cautiously supportive ('There used to be a free-for-all – thousands of seals were getting shot. Now at least there's some control'), while others expressed outrage at the fact that there's no independent monitoring of the culling and the shooters themselves are responsible for reporting the number of seals they kill. ('They can say any bloody thing they want – it's fairy stories. And the government is so far up the backside of the fishing industry that it smells of kippers.')

In the year after the Crovie tourists first raised concerns about the seal carcasses, George Pullar was quoted as saying, in the leaked minutes of a Salmon Net Association meeting, that 'it was pointless having a restriction on the number of seals which could be shot' and that 'a quota is not practical'.[3] After the story broke, he insisted that, although this was his personal view, Usan were continuing to adhere to all protocols and, by shooting seals under licence, simply taking the necessary steps to protect income and livelihoods.[4] Yet even while he worked to defend his company's reputation, a campaign to thwart the seal killing was getting underway, jointly orchestrated by the marine conservation charity, Sea Shepherd UK, and activists from the Hunt Saboteurs Association.

The HSA's Alfie Moon, though based in the south of England, was a prime mover in the incipient Seal Defence Campaign.[5] I managed to track him down before my current visit to Scotland and he readily spoke in detail and at length about his role.

'I've been a hunt saboteur for over thirty years,' he told me, referring to the many hundreds of weekends he's devoted to targeting fox hunts. 'Then in 2013 – this was before we were approached by Sea Shepherd – we started getting information about seal killings in Gardenstown. We came up with a plan to use some sea kayaks left over from an anti-nuclear warship protest that one of my friends said she could get her hands on, thinking we could launch them into the water between the shooters and the seals. I'm not sure how brilliant this plan was but it fell through, partly because the badger cull was starting up and people were committing to fighting that.'

The following spring, they devised another plan that was 'even crazier' than the kayaks. 'Just two of us from the HSA in wetsuits and flippers ready to jump in the water and put ourselves between the seals and the shooters. It sounds crazy but this is what we do when we sabotage the pheasant shoots – we literally put our bodies in the way. And if the shooter doesn't immediately remove the shell they've broken the law and will lose their gun licence.'

Sea Shepherd's approach and suggestion of a joint mission, however, added new layers to the plan, the first stage involving Alfie travelling to Gardenstown with Jenny Green, a fellow sab, to recce an extensive stretch of the coast and ascertain where Usan had set all their salmon nets. Now that I've seen the cliffs and coves of the Moray Firth for myself, I can better appreciate Alfie's references to the challenges of mapping the terrain and his relief at the 'stroke of luck' they received a couple of days after they got to Gardenstown and spotted an Usan boat coming into the harbour. 'Jenny, who's an absolute genius at this sort of thing, wandered up to them, fluttered her eyelashes and started talking. Within twenty minutes we knew everything about everything they were doing. Another ten minutes and we'd have known their mothers' maiden names and their sock sizes!' Having been convinced by Jenny's story that Alfie was her brother with a passion for wildlife photography, 'they told us about all these wonderful places we could go to and it was literally where all their nets were!' Over the next few days, equipped with this information, the sabs covered more than fifty miles of coastline and put together a map of all the net locations where the seal killings were likely to be taking place.

Although I also managed to make email contact with Jenny before my visit to Scotland, it would have been logistically tricky for us to chat as she was 'in the Middle East trying to get 3 baboons to safety'. Instead, an article she wrote for *Earth First!* journal and forwarded to me offered a useful insight into why the collaboration with Sea Shepherd worked so effectively. 'They have the boats, the marine wildlife knowledge, the media machine, and money. They are also a registered charity, which brings respect and a platform, but also means they cannot do certain things. . . We have a rawer modus operandi.'

Once Sea Shepherd UK arrived in Gardenstown, Alfie and Jenny handed over their maps to Chief Operating Officer, Rob Read, and withdrew, 'identities intact'. When I contacted Rob about the direct action phase of the campaign, he was especially keen to fill me in on the escalating tension. 'When we turned up, it got very, very heated – a lot of threats being thrown round, a few punctured tyres on cars, that kind of thing. We had a very simple principle – if Usan were out, we'd be following them. We'd follow their boat at a safe distance and if we couldn't stop them shooting seals we'd try to film or photograph them doing it.' On one occasion, he had to call the police from his boat as a mob around thirty strong was waiting on the harbourside at Gardenstown – 'a nice welcoming committee for us. They blocked up the entire harbour with their cars and were making threats from the pier to stop me bringing in my boat. A few incidents later, the police took more of an interest especially when the hunt saboteurs brought up a minibus load of people.'

'What about local residents?' I asked. 'Did they support and get involved with the campaign too?'

'Yes, some would tell us what Usan were up to. A few became our volunteers. And we'd get anonymous gifts – we'd come back from a patrol and there'd be a packet of biscuits sellotaped to the wing mirror of our van [with a note] saying "Thank you for defending the seals."'

In the following year, 2015, as a result of Usan having bought up hereditary salmon-netting rights to huge tranches of the north and east coast, Sea Shepherd expanded its focus to the Thurso area as well as Gardenstown while the HSA shifted its attention south to the Montrose region. 'We scouted a large area,' Jenny wrote. 'We set up observation points, mapped the nets, found the quickest routes to each net (over cliffs, down ravines, and over rocks), located safe places to leave equipment and vehicles. . . From April to September, a small team of hunt sabs held down ten coastal miles of enemy territory.'

Patrolling this ten-mile stretch of coast and Usan's fourteen salmon nets that were spread along it was relentless and physically demanding. Each net would be emptied and reset two or three times a day depending on tide times, and from 5 a.m., sabs would run along the coast between them, filming the process from the rocks at low tide or submerged in the water at high. As long as the cameras were trained on the fishermen they didn't risk taking a shot at a seal and, when a sab was on the scene, no animal was killed the whole season long.

In a rather unlikely alliance – again, according to Alfie, testament to Jenny's 'genius' – the support of the river anglers, a powerful lobby who resented the netsmen taking so many salmon along the coast, leaving fewer to be fished upriver, was also enlisted. Relationships were fostered, too, with a number of other organisations and individuals, including the Scottish Environment Protection Agency, the local community council and MP, and the Scottish Society for the Protection of Animals, who were able to offer advice and legal input and also lend authority to the hunt sabs' cause. As 'a rabble of activists from outside the country', Jenny acknowledged that the sabs' submitting of evidence against Usan could have been easily disregarded, but with the backing of salient organisations, due attention was paid.

Despite these positive developments, she admitted that they were regrettably unable to do round-the-clock monitoring of Usan's activities and six seals were shot in the Montrose region during the course of the season, including a heavily pregnant cow who aborted her pup on the beach as she bled to death.

It was, however, Sea Shepherd's discovery of a carcass of a young, flipper-tagged bull who had been shot near Thurso that roused the strongest public aversion to the slaying of seals. The Scottish Society for the Protection of Animals identified him as Kuiper, whom they'd rescued, rehabbed and released as a pup three years previously. 'That was pretty much game over – the media went crazy,' Rob told me. 'We had a video of [a marksman] shooting a seal and blood pumping out of its neck, still swimming away, but even that didn't have the same effect.' While it seems perverse that it took the killing of a named seal to inspire a wider outcry when the death of every animal should be of equal consequence, the attention that the story received in the media was much appreciated by those who were fighting to end the slaughter.

Meanwhile, around Montrose, Jenny and co. had filmed many incidences of Usan's nets being left in the water outside permitted fishing times: by law, for conservation reasons, no netting was allowed between 6 p.m. on Friday and 6 a.m. on Monday. The footage was submitted to, and investigated by, the relevant authorities, which led to a sizeable fine.

Over the course of the 2014 and 2015 seasons, the activists' presence undoubtedly resulted in a substantial reduction in the number of seal shootings in northeast Scotland. Yet it was the introduction of a three-year ban on coastal wild salmon netting by the Scottish government in an attempt to boost dwindling numbers of the fish that ultimately offered the seals longer-term relief. With the 2019 announcement that the ban was to be extended, Usan's setting of nets and attendant shooting of seals is on indefinite hold. All subsequent applications for seal-shooting licences linked to the creel pot side of their business have also been rejected.

Even though a number of years have passed since the end of the Seal Defence Campaign, Rob still sounded triumphant when we spoke – 'If we can win against Usan, we can win against anyone.' In my final email exchange with Jenny, however, a more circumspect note was sounded. While her *Earth First!* article emphasised the significance of the outcome, she alluded to the personal toll the campaign, and the prevailing atmosphere of menace and hostility, had taken. 'I was exhausted and depressed and broke . . . I knew I wouldn't win.

'So I disappeared.'

The next day, I decide to walk east from Crovie to the seabird cliffs of Troup Head, part-following in Jenny and Alfie's stealth reconnaissance footsteps. Compared with Alfie's description of their experience of the coast – a gust of wind that 'literally' lifted Jenny off her feet and a red deer stag that 'pinged up' in front of them as they worked their way through the gorse – my walk is quite uneventful. A cow seal surfaces just beyond the base of the cliffs and both wind and waves are subdued enough for me to hear her snorty exhalation when she prepares to dive again, but I still fail to spot any resident bottlenose dolphins.

The views of Troup Head's several thousand gannets, one of only two mainland colonies in the UK, more than compensate. From a belly-down position, I peek over the edge, pinned to the cliff by the piercing blue of their eyes, outlined in black like a goth's make-up. Breeding pairs clatter beaks, preen their young and let out raucous calls, heads and necks ambered like dawn. In flight, they're equally mesmerising, whether gliding or plunge-diving, folding their black-tipped wings behind them and slicing into the sea at a hundred kilometres an hour.

It took activists with the energy and focus of a gannet's plunge-dive to commit so wholeheartedly to the Seal Defence Campaign and, as I wriggle back from the cliff edge and sit cross-legged among tufts of grass and browned thrift flower heads, I think about my own much tamer experience of crusading. I've marched, signed petitions, boycotted palm oil, ditched all single-use plastic. I joined the World Wildlife Fund as a Junior Member and saved up weeks of pocket money to adopt a panda. I wore my Body Shop *Against Animal Testing* T-shirt till it fell into tatters and I've sheered between vegetarianism and veganism for most of my life. But in comparison with the direct action of Jenny and co., my version of campaigning is no more than Activism Lite.

As a child, my eyes were opened to the need for animal advocacy by the emotive photo in Dad's encyclopaedia of a harp seal pup, purity-white, lying helpless on the ice, fated to be clubbed and skinned by Canadian hunters. Later, I blutacked Greenpeace protest posters to my bedroom wall and read all I could stomach about what was happening. In the nineteenth century, millions of harp seals were killed in the Gulf of St Lawrence and at the Front, the ice-field off the northeastern coast of Newfoundland, as their oil lit city streets, was a constituent of soap, and lubricated machinery. In the twentieth century, however, the hunt was driven not by oil but the fur trade – the pelts of harp seal whitecoats and young hooded seals, known as bluebacks, were highly coveted in both the USA and Europe. Campaigns from the 1970s onwards included activists spraying the pups with organic dye to destroy the commercial value of their pelts, to recruiting celebrities such as Brigitte Bardot to the ice to help raise global awareness. Eventually, in the early 1980s, the European Economic Community (EEC) voted to ban the import of pup pelts and in 1985, my letter would have been one of the 60,000 sent to the UK government to urge it to support an upholding of the ban. Yet, in spite of the fact that the demand for seal products has fallen due to the closure of many international markets, the Canadian government has persisted in financially supporting the industry and setting a quota that has recently permitted up to 400,000 harp seals to be killed each year. While the hunt currently seems to be on a declining trend, it still remains the biggest government-subsidised wildlife slaughter on the planet.

It wasn't until much later, when I lived and travelled in Canada, that I became aware of the fact that its maritime provinces are also home to grey seals, rather larger versions of those in British waters as the Northwest and Northeast Atlantic populations have been reproductively isolated since the last glaciation and evolved some morphological and behavioural differences. Having since read Linda Pannozzo's passionately argued *The Devil and the Deep Blue Sea: An Investigation into the Scapegoating of Canada's Grey Seal*, I'm conscious that the attitude and strategy of the Canadian fishing industry and politicians towards the species have often paralleled those of their UK counterparts. Following what some biologists maintained was its near-extirpation by the end of the 1940s, continued culls and a bounty programme that paid fifty dollars to licensed fishermen who could offer up an adult's lower jawbone as proof of its demise, the grey seal population was, more recently, permitted to rebound. However, this predictably spawned a rancorous debate around the impact of their presence on fish, and cod numbers in particular.

Travelling through Newfoundland, I listened, on many occasions, often in cafés that served seal flipper pie, to febrile discussions about the cod moratorium – the fact that cod stocks on the Grand Banks had collapsed after decades of over-exploitation, resulting in the Canadian government issuing a fishing ban in 1992. Thousands of Newfoundlanders consequently lost their jobs, unemployment soared and many left the province to seek work on the Canadian mainland. As I travelled, I also became aware of former fishermen diversifying, particularly into the tourist industry – setting up B&Bs; foraging for partridgeberries, bakeapples and crowberries with which to make and sell jam; performing in dinner-theatre shows; reinventing themselves as folk artists. Some had undoubtedly made a success of these alternative enterprises, but, years on from the moratorium, there was still considerable lingering bitterness. Though I sympathised with many of the ex-fishermen I met, I also felt troubled by the fact that so much of their bitterness was directed towards the grey seal.

Thorough historical and scientific research has been conducted to prove why this scapegoating is unwarranted. Prior to, and in the early years after, European contact, it's been estimated that the grey seal population was up to three times higher than it is today and yet the fish population was equally plentiful: a basket lowered over the side of a boat into the sea in the fifteenth century would be drawn up overflowing with cod. Scientists have discovered, too, that cod isn't the staple food of the grey seals breeding on Sable Island but that they prefer herring, capelin and sand lance due to the fact that they have a higher fat content. Since these fish themselves feed on cod larvae, it could be argued that seals are actually helping cod numbers to recover by eating their predators. It would also be judicious to consider other possible explanations for cod's slow recovery such as the high concentration of toxic chemicals in the Gulf of St Lawrence from the industrialised Great Lakes region, ocean acidification, noise disturbance from seismic testing for oil and gas reserves, and trawling and dredging activities that damage seabed habitats.

The first decades of the twenty-first century have seen several further grey seal culls, with many at provincial and federal government level persisting in regarding the species either as pest or source of profit – there remains an Asian market for seal penises, for example, sold as sexual enhancement products. Meanwhile, anti-cull campaigners continue to question how, in the words of fisheries scientist Jeffrey Hutchings, we can ethically condone 'the deliberate killing of one native species for the sole purpose of possibly increasing the economic gain obtained from harvesting another'.[6]

This question is very much at the forefront of my mind as I leave the clangour and clamour of the gannets at Troup Head and retrace my steps to Crovie. As well as a bottling bull seal, snout raised skywards, I at last get a distant glimpse of a small pod of bottlenose dolphins, their fins and flanks emerging and submerging like a needle sewing a running stitch. Yet, wildlife aside, what I'm finding most striking about today's views is the fact that they contain none of the huge floating circular or square structures, beneath which underwater nets hang, that I grew so used to seeing on my way up the west coast to North Uist and around the isles of Orkney. These pens contain farmed salmon – though there may not be any here on the east coast, more than two hundred aquaculture sites operate elsewhere in Scotland, making it the third-largest producer of farmed salmon in the world. And while there's no longer a seal-shooting issue at wild salmon nets as a result of that fishing practice having been banned, killings still occur around fish farms.

Many other animal welfare and environmental concerns, in addition to seal shootings, have been voiced in respect of intensive salmon farms in recent years. These include overcrowding – each pen holds many thousands of fish with a percentage struggling to feed and starving to death; infestations by parasitical sea lice, in part due to the congested conditions, that can eat the salmon alive; the use of toxic chemicals, with implications for the wider marine environment, to attempt to control the sea lice; unsustainable harvesting of wild wrasse that get dumped into the pens to act as 'cleaners' as an alternative means of parasite elimination; and the use of thermolicers, heated water chambers, into which the salmon are sucked, which have at times resulted in accidental mass mortality events rather than successful de-licing.

That seals are attracted to pens containing thousands of live, and a proportion of dead, fish is no surprise, nor is the salmon industry's claim that seal attacks cause serious losses of both fish and equipment. It insists, too, that seals are only shot under licence by skilled marksmen as a last resort but I remain perplexed by this argument, the legitimisation of killing that the licence scheme seems to offer and the fact that there's no external monitoring of the number of shootings or an assessment as to whether quotas are being adhered to or exceeded. Eager to discuss these issues, I've made contact with the Seal Protection Action Group, whose long-standing Saving Scotland's Seals campaign contributed, in part, to the Scottish government's decision to introduce the licensing scheme on the basis that it would outlaw unregulated shooting.

In a phone call that stretches much longer than expected, like the foraging journeys of weaned pups, the director of the Seal Protection Action Group, Andy Ottaway, explains that it grew out of a small organisation which was formed in Orkney in opposition to the 1970s cull. Now a charity with a national profile, its mission to safeguard seals and their environment is fulfilled through lobbying for effective legislation, running public education campaigns and financially supporting mobile rehab units for injured and orphaned pups. Andy's a veteran ocean wildlife advocate, having worked in marine mammal protection for more than thirty years. 'Throughout my career, I've felt that seals are pretty much second-class marine mammals – no one seems much concerned. So many groups have focused on the popularity of whales and dolphins – I fully understand their iconic nature, fully understand that they need protection, but there's sort of a whale mafia.'[7]

Given his undoubted long-term commitment to improving the welfare of seals, I'm wondering about the degree to which he can offer a positive slant on a licensing scheme that permits them to be shot for the protection of fisheries and aquaculture activities. Killings are even permitted during the breeding season, meaning that a pup will also die if its lactating mother gets shot. In response to my reservations, Andy concedes that the licensing scheme isn't ideal by any means but it's still 'vastly better' than the 'unrestrained and unregulated persecution of seals' that was taking place before.

'But how do you know that rogue killings aren't still taking place as well as licensed ones?' I ask.

'I think what we've got now is quite an interesting dynamic,' he says. 'Those who are abiding by the law don't like those that aren't. You create an environment that's almost self-policing. The last thing the aquaculture companies need is a rotten apple who's shooting seals and not reporting them and bringing the whole thing into disrepute. And it seems to have gone from a situation where we thought thousands of seals were being shot each year to maybe a couple of hundred. When the licence scheme first began, people were asking for huge numbers of seals to be shot but those requests have gone down and down. Now they have to count the bullets so to speak, it's brought in some restraint.'

He admits, however, that the Seal Protection Action Group's pragmatic approach hasn't been universally popular with other campaigning organisations. 'We've also opened up a dialogue with the leading producer of salmon and the leading retailers, and suffice it to say that some of the more rabid amongst us accused us of being traitors and selling out. You won't find anyone more abso-lutist than me when it comes to the persecution of animals – our bottom-line position is that all seal shooting should stop – but I have to be a pragmatist too. They're not going to close down a half billion pound industry overnight because someone's telling them to because they're shooting seals. Staying on the far wing wasn't going to achieve anything – we weren't even going to get a conversation.'

The dialogue that Andy opened up led, after much negotiation, to the establishment of the Salmon, Aquaculture and Seals Working Group. Comprising government representatives and salmon farmers, animal welfare groups, retailers and scientists, the group meets several times a year to explore non-lethal means of reducing seal predation on salmon that should obviate the need for any shootings. 'We didn't want to create a waffle shop or some kind of fig leaf of respectability that the industry could hide behind,' Andy insists. 'And some of it's successful and some of it's dead end, but that goes with the territory.'

Though the industry has experimented with the installation of acoustic deterrent, and louder acoustic harassment, devices, some of which emit continuous pulses of noise while others are activated by the agitated movements of fish in response to a seal attack, they are said to impact negatively on cetaceans, causing hearing damage, stress and displacement, and masking the sounds produced for communication and prey detection. Seals may also become habituated, learning to associate the noise with the presence of thousands of fish and responding as if it were the ringing of a dinner bell.

More investment is warranted, too, in alternative netting systems that keep salmon and seals apart – double netting and/or high-density polyethylene nets that are much tougher and more rigid than traditional nylon. Additional weights at the base of the nets to increase tension and make it even harder for seals to penetrate them are also recommended. Those aquaculture sites that have so far been willing to invest in the next-generation nets have reported less predation and a reduction in seal shootings as a result. An 'electric fish' deterrent has also recently undergone trials – placed at the base of the net, it gives a minor electric shock to seals that attempt to grab it, dissuading them from persisting any further.

Elsewhere in the world, the aquaculture industry has employed other anti-predator methods including the translocation of seals that are repeat offenders, the deployment of fibreglass models of predatory orcas, and conditioned taste aversion, a process by which an animal learns to avoid a site where food's been doctored to make it vomit. No such experiments have been attempted in Scotland, though several decades ago an informal test of the response of seals to specific repellents was carried out at a west coast fish farm. Salmon laced with chilli and curry powder was suspended from the pens but seals apparently approved of the taste and weren't discouraged from eating it.

As for other means by which the aquaculture industry might be influenced to terminate the shooting of seals, Andy confesses to being 'a bit of an old dinosaur – I believe in consumer pressure. If the retailers say we don't want this, our customers don't like seals getting shot, the producers will have to change their practices. But to mobilise huge numbers of people isn't easy and it's not easy to sustain it either.'

Part of the problem lies in the fact that if a salmon product bears an RSPCA Assured label, consumers will assume that the highest animal welfare standards have been met at the site where it was farmed. However, while the salmon may have been cared for in a suitable manner, seals are still shot under licence at farms that have been assessed and approved by the RSPCA.[8] 'Their position is frankly ludicrous,' Andy says. A 2018 Twitter spat in which the RSPCA's head of campaigns and public affairs declared that 'Seal shooting is not culling, it's about humane pest control'[9] also exposed the inconsistency of the organisation's approach, performing high-profile pup rescues and pumping funds into rehab at their sanctuaries on the one hand and approving the shooting of 'problem' adults on the other.

As our conversation starts to wind down and my mind turns to the next call I need to make, my usual touching-base chat with Dad, Andy reveals that a new strand has been added to the snarl of ethical and financial complexities that characterise the Scottish aquaculture industry. It appears to offer the most hopeful route yet to an end to seal shooting and comes from an unexpected source. In the United States, new regulations, building on earlier legislation in the Marine Mammal Protection Act, will result in seafood products being banned if the importing countries fail to match the standards of marine mammal welfare that are upheld by the US. While the measures, due to come into effect in 2022, continue to aim to reduce the number of creatures killed as bycatch in commercial fisheries, they also extend to the protection of marine mammals at fish farms. Salmon is Scotland's top food export and the US market is worth well over £100 million a year but unless seal shootings are prohibited and rigorous monitoring procedures put in place within the designated time frame, it seems that Scotland will be barred from accessing that market.

'We'll keep pressing on this issue and see what happens,' Andy concludes. 'Money talks. Very loudly.'[10]

14

Stampede

The film's beginning is beautiful and benign. An aerial tracking shot of billowing dunes edged by a dazzle of sea. Twin curves of sand, full lips framing the mouth of an estuary. Suddenly, though, it shifts in genre, ditching the tropes of *Coast* for horror. The camera swoops low and what looked like rocks on the bottom curve of sand morphs into hauled-out seals, their heads jerking up in alarm at the drone's arrival. Lower still and their bodies jerk into action too – close to a thousand of them scrambling over each other, trampling some of the younger ones in their panic.

Over the course of yesterday evening – before, after and even during my phone call with Dad – I watched numerous versions of such disturbance incidents, as suffered by the seals of the Ythan Estuary just north of Aberdeen. Footage of both air- and land-based incursions has been posted to the Facebook page of Ythan Seal Watch by Lee Watson, the volunteer group's founder and co-ordinator, as a means of raising awareness of the seals' struggles and support for his campaigning work. I've pootled a few hours south of Crovie via the scenic coastal route and am due to meet Lee this morning for an insight into one of the largest grey seal haul-outs in the whole of the UK.

A brown 'Seals' road sign, positioned beneath two others for 'Golf Club' and 'Beach', points me to a car park between fairway and gorse, where there's another compilation of signs. 'To get the best view of the SEALS please follow this route.' 'SEALS ahead. Please keep dogs under close control to avoid potential injury to your dog.' And, as a counter to all the chivvying, a jaunty info board peppered with exclamation marks. 'Grey seals make lots of noise! Listen to them singing and moaning at each other!' 'Seals moult their fur and in spring especially, you may see them scratching and wriggling on the sand to cure an itch!' 'At only three weeks old, their mums leave and they must learn to survive on their own!'

As I wait for Lee, the board gets peppered with hailstones too, a brief but fierce interlude between spells of breezy sun.

'Sorry about the weather.' When he arrives, along with his wife, Eilidh, he seems deeply embarrassed that the elements have given me such an unseasonal welcome. 'Very sorry.'

We pick up the path through the gorse past a dilapidated fishing shed, such a contrast to the swish golf course clubhouse overlooking our route. Now that the hail has abated, we all take down our hoods – Lee to reveal gingerish hair and sideburns and Eilidh a ponytail of purple and pink. After a few minutes, we emerge on the south side of the Ythan Estuary onto an amalgam of sand and mud and lugworm casts. The tide's fairly low at the moment and the rotting hull of an upturned boat is exposed in the centre of the channel through which the river flows. Opposite, on the north bank, is the dune system that featured so prominently in all the disturbance incident videos.

'That's Forvie National Nature Reserve,' Lee tells me, 'managed by Scottish Natural Heritage.'[1] The seals haul out on the beach in front of the dunes – we'll have to walk on a little further before we'll be able to see round the corner of the estuary mouth and observe them. While only drone pilots can disturb the seals from our side of the river, around thirty-five thousand hikers, dog walkers, photographers and birders explore the reserve on the other side each year and though a fence has been erected to encourage them not to access the seal beach, some refuse to be deterred. Lee has been monitoring the haul-out since 2015 and he set up Ythan Seal Watch and started using the Facebook page to inform and educate about the impact of disturbance the following year when the seals were especially stressed.

'If we weren't there by ten in the morning, the whole haul-out would be gone. We knew it wasn't sustainable – the beach would be overrun with visitors to the point that the seals weren't able to access it at all.' Though the situation has improved somewhat since then, he might still witness, and often captures on film, four or five disturbance incidents each weekend. 'This is what I use – just a compact camera, just a little wee thing,' he says, reaching into one of the pockets of his waterproof. 'There are two kinds of disturbance – naive ones, people who see others over there and think they'll go over too, and those who delib-erately disturb. We can tell the difference by their behaviour – the ones who sneak up and duck down know exactly what they're doing. Some take photos of the stampede and sell it to Shutterstock. Some give us abuse, some give us the finger. There are genuine mistakes though too – we once went over to the car park on the other side to tell a woman about the damage she'd done and she started crying.'

He's well into his narrative stride now and is clearly ready to describe many more of the struggles and successes experienced by Ythan Seal Watch. Eilidh, though, is itching to chip in with information about some of the other wildlife that frequents the estuary. Our conversation's had a rowdy backing track – the shrieks and squawks of hundreds of Arctic, Sandwich and common terns from their protected nesting area. The breeding population of the three species at Forvie is of international importance, as is that of the dainty little tern, which has been in serious decline in the UK in recent decades. Eilidh scrolls through the images on her camera – it's a lot more sophisticated than Lee's 'little wee thing' – to show me some of her little-terns-in-acrobatic-flight shots and confesses that her interest in, and concern for, seals evolved from her passion for wildlife photography.

We resume our amble towards the estuary mouth, accompanied by the river and several other visitors. A willowy adolescent kicks off her UGG boots and rolls up her jeans, then grimaces when her feet sink in the sludgy sand. Her father's preoccupied with their miniature schnauzer who's scampering round him in giant circles, manically digging, then raising her head to yap with sand caked in her beard.

The fact that she's on a lead gets a Lee Watson nod of approval. Dogs often cause stampedes on the Forvie side, he says, when their owners let them run free. A dog that's encouraged to swim can quickly experience difficulties too – when seals are in the water, they're much more confident than when they're on land and if curiosity prods them to gather round a dog, it may not be able to access the beach again and will instead be carried out to sea on the strong Ythan current.

We've reached a line of tripods – about twenty of them, behind each of which a photographer is stooping – plus a tutor-guide making suggestions. Following the direction of their mega lenses, I realise it's now possible to see to the beach beyond the estuary mouth and my gaze steadily pans between one, two, three, four large groups of seals. With the aid of binoculars, I estimate that each group contains several hundred animals, a mix of adult cows and bulls and a large proportion of youngsters. First-year juveniles must be especially vulnerable to the effects of disturbance – as well as being more likely to get injured or crushed to death in a stampede, they're also still learning to feed themselves, so the energetic cost of being flushed into the sea when they're trying to rest is all the greater.

'Our peak numbers are in winter and early spring,' says Lee. 'The conservative estimate is that we get around two thousand seals but it's been up to three and four at that time.'

'We continue to get big haul-outs throughout the year though,' Eilidh adds.

I've been wondering if there's a seasonal element to their monitoring, a period when they don't have to be so vigilant, but no, it's mostly been an all-weekend-every-weekend commitment. When they're not working at their day jobs – Lee for BT and Eilidh at a doctor's surgery – they're devoting the bulk of their free time to trying to protect the seals' welfare.

Some weekends, Lee says, he bases himself on the other side of the estuary so as to intercept all those people who are making a beeline for the seals. 'By the time I've finished, I'm exhausted – I'll have walked about twenty miles and I still have to keep reserves in my legs for getting back to the car park.'

'Do you have a team of volunteers you can draw on?' I ask.

He shakes his head. 'No, but if a seal hauls out on this side of the estuary, as the young ones do sometimes, we'll get a phone call and people will watch it for a few hours and tell other visitors to stay away from it. Members of the public send us photos and videos of disturbance for our archives too. But otherwise, it's just me and Eilidh doing it – we're not a charity, there's no investment, nothing.'

We're on the beach proper now, the willowy teen visibly relieved to be free of the estuary mud. A couple of children are playing hide and seek among a row of concrete blocks, installed, Lee tells me, as anti-tank defences in the Second World War when the beach was deemed to be at high risk of invasion from German units in Norway. It's delightful to see young seals on the Forvie side like-wise playing around an old military construction – launching themselves in and out of the windows of a pillbox, half buried in the sand.

Even though Lee admits that he sometimes questions 'whether it's all worth it', the Ythan Seal Watch cause was ostensibly given a major boost in 2017. The Scottish government passed an amendment order to the Marine (Scotland) Act 2010 which saw the Ythan be included in its Designation of Haul-Out Sites. This order is meant to legally protect the seals from 'intentional and reckless harassment' and allow them to rest, moult and breed undisturbed. Disrupting a designated haul-out in Scotland is classed as a wildlife crime and carries a fine of up to £5,000 or six months in prison.

Straightforward though this sounds in theory, the reality has proved to be much less clear-cut. Marine Scotland, a directorate of the Scottish government, offers guidelines for, rather than an explicit definition of, 'intentional' harassment, stating that 'there is considerable case law on what constitutes intent and ultimately only a court can judicially determine' whether or not an action was deliberate. While there is also 'considerable case law on what constitutes harassment', this pertains to people rather than wildlife. It outlines various activities that could fall under the harassment banner, one of which, in particular, perfectly describes disturbance incidents at the Ythan – 'Intentionally or recklessly approaching or sneaking up on seals on designated haul-outs from the landward side' – but insists that 'the guidance is advisory only' and not legally binding.

Lee believes that a key factor is whether the same person repeatedly offends and, on multiple occasions on the Ythan Seal Watch Facebook page, he quotes from a Scottish government list of FAQs, some of which aim to expand on the harassment guidelines. 'It will only rise to the level of harassment either where the majority of seals are massively disturbed in a single incident or, much more likely, where large numbers of seals are significantly disturbed on a repeated or on-going basis.' Lee's experience of the police's approach seems to reflect this governmental stance. 'I've never spoken to a wildlife crime officer – never in all this time. They know what's happening but they're not going to do anything unless we catch the same guy twice.'

As a result, he's now painstakingly checking back through the hundreds of images and videos in the Ythan Seal Watch disturbance archives from the date of the designation onwards to try and work out if any visitor has intentionally sparked off a stampede on more than one occasion. He's also studying drone footage, of the kind by which I was horrified yesterday evening, that pilots post on YouTube. 'They deliberately buzz the seals to cause a stampede, then laugh about it online.'

While we've been talking, the tide's started snaking in, sneaking up on us from behind. The tripods have moved further up the beach while a couple of juvenile seals on the opposite side are turning their loss of haul-out space into a game, struggling onto a higher, tide-free ledge of sand, then letting themselves fall back into the water.

'A few months ago, it looked like they made a slide,' says Eilidh. 'They kept climbing up onto one of the dunes and sliding down, one after the other.'

While Lee uses his binoculars to scan the sand for footprints – the haul-out's officially closed from April to August as reaching it means invading the protected ternery, yet people still find ways to get there – Eilidh offers a commentary on the young seals' antics. This site, with its high proportion of juveniles and enchanting play behaviour, is quite different from any other I've been to.

'Were many of the young ones born here?' I ask.

'No, they come from other breeding colonies – the Isle of May off the Fife coast is the nearest and then there's Orkney too,' Eilidh tells me, adding that here, there tends to be just one pup per year, with the same cow having given birth several seasons in a row. 'And oh, the stress of watching that poor wee thing! I hope the other cows won't start – I wouldn't be able to cope with that.'

One year, Lee laments, the just-weaned pup was caught up in three stampedes in the course of a week – panicked by a drone, almost flushed off the beach while still an inexperienced swimmer by a visitor who jumped the barrier fence with an iPad, and frightened by a dog. It was lucky to survive.

After hearing this worrisome tale, I feel aggrieved to come upon a woman who's letting Alan the spaniel gallop alongside the estuary and dart in and out of the water. When Lee politely suggests it would be wiser to put him on the lead as the tide's coming in and bringing some of the seals with it, she does so but is markedly miffed. 'One dog is fine around a seal,' she snaps. 'Only a pack would be problematic.'

The estuary mouth is gargling with the rising water and an increasing number of young seals are plunging through the chop and churn. We turn to walk alongside the Ythan again, back in the direction of the car park, so as to watch the seals who've let themselves be washed off the beach and out of the pillbox come surfing upriver. Those who are choosing to stay on the narrowing stretch of sand have started moaning and howling as if venting their displeasure at the sea. Other wildlife seems to be similarly animated by the incoming tide – over towards the breeding terns, there's a commotion of black-headed gulls. 'We'll often see a peregrine in the middle of them,' says Eilidh. 'And sometimes an osprey comes to feed here too,' though binoculars confirm that's not the case today. Just ahead of us, a seal surfaces and flusters a cluster of eiders. I'm stunned by the speed at which this bulky duck can fly – I've previously only ever seen them swimming or lurching along on land yet, as far as level flight's concerned, they've been adjudged to be one of the world's fastest birds.

It strikes me that Lee and Eilidh have had to show equivalent momentum and intent in their work with Ythan Seal Watch over the years. Even now, Lee's outlining some of the developments that they're going to campaign for in the months ahead including, as a deterrent to people thinking of approaching the seals, signage advising of the possibility of prosecution.

A few of the bolder and more curious youngsters are swimming close to our side of the river, looking as if they might consider clambering out. One is porpoising against the flow while others are letting themselves be carried towards the algae-coated hull of the rotting boat. With the water swirling around it like a huge jacuzzi, it provides another opportunity for group play. Unfortunately, the instinct to interact with new, unfamiliar objects has stretched to one seal investigating a large plastic bag – she's swimming along with it trailing from her mouth, diving then re-emerging, still with the handles between her teeth

'There's something else we're going to be doing too,' says Lee. Following contact with Sue Sayer, who, with her experience of setting up and co-ordinating Cornwall Seal Group Research Trust, has always been ready to offer advice to him and Eilidh, they've decided to tweak the Ythan Seal Watch Facebook page. 'We need to get as many folks on our side as possible so we're not just going to make angry videos of people causing disturbance any more.'

Thanks to the high jinks of juveniles that seem to characterise this special site, it should be an easy task to follow Sue's recommendation. 'It's about time you guys had a bit less vinegar on your page,' she said. 'And a bit more honey.'

Orange clownfish. Bubble-tip anemone. Skunk cleaner shrimp. Pink scooter blenny. A shoal of academics is moving through the foyer of the Scottish Oceans Institute at the University of St Andrews where the Sea Mammal Research Unit is based but I'm more mesmerised by the inhabitants of the display tanks near the door.

I'm on my way south to the Isle of May, flagged up by Eilidh as the site of the nearest seal-breeding colony to the Ythan, where many of the haul-out's frisky juveniles could have originated. Since my route takes in the town of St Andrews and I've heard mention of the Sea Mammal Research Unit on several occasions during the course of my seal journey, I've managed to arrange a meeting with Professor Ailsa Hall, its director. She's very pressed for time so I'm grateful to have been granted an hour in which to learn more about SMRU's role and some of the latest grey seal-related research by its scientists.

While the building's exterior surrounds – rigid inflatable boat plus trailer, view of the sea and a woman swapping her heels for windsurfing boots – sets it apart from all other university premises I've known, Ailsa shows me into a generic office for our chat. With her tousled fair hair, cropped blue trousers and pumps, though, she looks like she'd be happier if we were meeting on a leisure craft than at a table flanked by a pair of computers.

'The original reason for SMRU's existence – and I make no bones about it – was because of seal–fisheries interaction,' she begins, getting the sensitive issue of its supplying of data to the government via the Natural Environment Research Council at the time of the 1970s Orkney cull out of the way early in our conversation. 'Now we're all about three things – discovery, the technology that enables the discovery and policy-related science.' She picks up a pen and waves it around as she gives a broad overview of some of her colleagues' grey seal research projects. Diet. Foraging. Physiological studies. Population modelling. Using new mobile phone technology to track seals. Analysing dive data. 'Looking at reproductive success of known individual females on the Isle of May – seeing who's still around, who's had pups, how the pups are faring. And relating that to environmental conditions, climate change, sandeel availability. . .' The pen comes to rest on the table. 'And we also have the captive facility out the back.'

This must be where the pups who were taught to sing 'Twinkle, Twinkle, Little Star' were kept when Vincent Janik and Amanda Stansbury were studying the vocal capabilities of grey seals.

Ailsa reveals that several young seals are again in residence but, anticipating what my next question might be, quickly adds, 'We can't take visitors there as the animals are so inquisitive and it'll disrupt their training.'

Instead, she switches to talking about the policy aspects of SMRU's work, which includes the pursuit of research into such areas as human impacts on marine mammals, and the provision of evidence-based science to those who develop national and international marine environmental strategies. In addition, in recognition of the fact that, under the 1970 Conservation of Seals Act, the Natural Environment Research Council is still obliged, post-Orkney cull controversy, to provide advice to the government on issues relating to the management of seal populations, a Special Committee on Seals has been appointed. SMRU annually offers scientific recommendations to this committee in response to questions on a broad range of seal management matters tabled by national government and the devolved administrations via DEFRA, Marine Scotland and Natural Resources Wales.

'Management' is a word of which I've often been a bit suspicious when it's applied to wildlife, having heard about several schemes from years past that manipulated 'resources' and 'stocks' according to human need and neglected to adopt a more biocentric strategy. Nevertheless, I try to set my unease with the term aside and focus on what Ailsa's telling me. I've waded through all the annual reports produced by the Special Committee on Seals, and some of the tabled questions on, for example, fisheries bycatch and interactions with renewable energy developments relate to issues that have been troubling me while I've been travelling, so it's going to be fascinating to get her perspective on them too.

To begin with, we discuss bycatch, an appreciable problem for seals, with one of the hotspots being the waters off southwest England, as I remember from my encounter with Sue Sayer. The estimated number of bycaught seals fluctuates but is generally believed to be between four hundred and six hundred per year. While this is not a large percentage of the total UK population, approximately 85 per cent is thought to occur in the Celtic Sea and western end of the English Channel, meaning the smaller, local, grey seal populations in those regions are disproportionately afflicted. The figure also fails to take into account bycatches in Irish, French and Spanish gillnets in the same areas, the extent of which is unknown, meaning the scale of the issue is likely to have been underestimated.

Ailsa acknowledges that entanglement in lost or abandoned fishing gear and other forms of ocean litter is a significant problem too, but when I mention the juvenile with the plastic bag at the Ythan, she asserts that deliberate consumption of marine debris that can cause intestinal injury and blockage is not such a crucial issue for seals. 'I think they're quite good at avoiding it. They're not as dumb as turtles – they can tell the difference between a plastic bag and a jellyfish and discriminate in many cases between what's edible and what's not.' I presume that accidental or indirect ingestion of microplastics – fragments and fibres that are less than five millimetres in size – via already-contaminated prey, could potentially be much more consequential. However, while studies from a number of academic institutions have investigated microplastics abundance through gut analysis of stranded, and scat analysis of captive, animals, including residents at the Cornish Seal Sanctuary, understanding of the impact on seals' health and population levels is at a very early stage. It's possible, though, that while microplastics may be readily expelled by seals, they can still linger in the stomach prior to excretion and this could enable the discharge of harmful chemicals.

When I move on to talk about the tidal turbine in Ramsey Sound and how conflicted I felt by its lurking presence on the sea floor when I visited Ramsey Island, Ailsa sympathises with my ambivalence but is reluctant to be drawn into expressing an opinion on the ethics of marine renewable energy developments. 'Work's being done on collision risk – how likely seals are to die if they get hit by a tidal turbine blade, how likely they are to avoid it, how they detect it and how we can deter them from it,' she says, tracing turbine-like circles in the air with her pen again. 'And it's possible that, at certain times of the year, certain age classes could be more at risk.'

I nod in understanding, thinking of a report I recently unearthed, which confirmed the tendency of some of Ramsey's weaners, a sample of whom were tagged, and tracked on dispersal from the breeding beaches, to favour the sound's tidal rapids and perform what appeared to be foraging dives.

But as for whether 'it's better to kill ten grey seals or save the planet,' she continues, posing her own provocative question, 'as a scientist, I don't have to answer that – that's a policy decision. Without my scientist's hat on, I'd say it's about compromise. We need to maintain a balance, have a sensible understanding of what the risks and our responsibilities are. But it's pretty complicated – a huge dilemma – so I don't want to have to answer that question and I'm not going to!'

Despite the assorted threats to grey seals' welfare that we address over the course of our meeting, it becomes clear, by the end of our allotted hour, that she's a good deal more optimistic about the resilience of the species than any ranger, warden, sanctuary manager or animal care worker with whom I've spoken so far. 'Grey seals have been resilient to a lot of insults over the years,' she says, and in the next fifty to a hundred, she thinks they'll be much more successful than commons 'in spite of things like fisheries interactions, as they're more adaptable and catholic in their tastes. Though they mostly eat sandeels, I suspect they have sufficient plasticity in their habits to say okay, we'll switch to cephalopods if the sandeels, for example, are no longer around.'

Aside from variations in the availability of their prey, I'm keen to know, though there's little time left in which to discuss it, if Ailsa's concerned about other ways in which climate change may negatively impact on grey seals. I mention increased storm severity of the kind with which I've become familiar in Pembrokeshire, but she's more bothered about the fact that climate change will lead to the emergence of new viruses and bacteria. After specialising, in the early years of her career, in human epidemiology, she diverged into marine mammals and an important portion of her work has since focused on the immune and endocrine systems of grey and common seals, including the impact of their exposure to toxic contaminants. Much of her field research has been conducted on the Isle of May alongside a small colony of fellow scientists and a large colony of breeding grey seals.

Compared with the 'poor old' common seal, 'the good old grey is like the rat of the sea,' she says, referring to the common's weaker immune system and susceptibility to the phocine distemper virus of which the grey is predominantly an asymptomatic carrier. 'But,' she adds, as the hand that waved the pen now shows me to the door, 'that's not to say we shouldn't be mindful of threats. The potential for new diseases as a result of climate change can't be dismissed. And if the phocine distemper virus mutates, the grey may no longer be resistant. We must never be complacent.'

15

Deconstructing Blubber

'Aye,' says one of the two crew members of the *May Princess*, 'summer's on a Tuesday this year.' Though we've set sail into no more than a faint breeze and the surface of the water is only mildly dishevelled, he still can't resist cracking a seasickness joke. 'If anyone feels unwell at any time, let me know,' he says, brandishing a single-use plastic bag. 'Just got to remember it'll cost you 5p.'

From St Andrews, it took me less than half an hour to get to Anstruther, the largest of a string of historic fishing villages along the north coast of the Firth of Forth in Fife. It's only just past midday yet there are already customers waiting outside the multi-award-winning Anstruther Fish Bar, billed as the most famous chip shop in the country. The queue extends past Chest Heart & Stroke Scotland, the charity shop sounding a caution from next door.

A number of my co-passengers have brought epic portions of chips on board to sustain them through our hour-long voyage to the Isle of May. I'm sitting on the top deck of the little boat, sandwiched between a man who's optimistically shirtless, his torso flecked with midge bites, and a woman who's more cautiously clad in a windcheater lined with fleece. On the bench opposite, a guy is reclining with his head on his girlfriend's lap: he's taken off his walking boots and socks and every so often grabs a grubby foot to pick at the calluses.

The Isle of May lies five miles offshore at around the point where the Firth of Forth segues into the North Sea. Owned and managed as a nature reserve by Scottish Natural Heritage, it looks to be a sawtoothed mix of cliffs and clefts, overseen by an unusual, castellated lighthouse. It's been suggested that its name derives from an Old Norse phrase that roughly translates as Gull Island, but as we cruise closer to one of the cliff walls, splattered with white guano like a Jackson Pollock, it strikes me that Guillemot Island would be far more appropriate. These dark brown and white auks are huddled in their hundreds on the narrowest of ledges – around sixteen thousand pairs nest on the island, we're told. Though it's more likely to be an evening spectacle, I keep my eye out for jumplings, as the chicks are known. At this stage in the breeding season, the adult males are likely to be gathering at the base of the cliffs to summon their offspring down. Even though its wings are not fully formed and it's unable to fly, each chick makes an audacious leap from the ledge where it hatched three weeks previously into the water way below. If it survives and successfully reunites with its father, it will be led out to sea and cared for until it's well-enough developed to be independent. During that time, a few, I presume, will be snatched by grey seals, who have been observed opportunistically snacking on seabirds that were swimming at the surface in a number of UK locations.

Extraordinary though it must be to witness the guillemots' premature fledging, the majority of my fellow passengers are more eager to see some of the island's forty thousand pairs of breeding puffins. The alcove inside the boat that doubles as a gift shop and bar is doing a great trade in the same puffin keyrings and magnets that I recognise from trips to Skomer. And when we reach the landing jetty to be met by the reserve manager, David Steel, a cheer goes up as soon as he mentions that puffin chicks are starting to fledge at night, walking from their burrows to the sea after spending the first forty days of their lives underground.

As part of our island induction, he goes on to speak about the other seabird species, from kittiwakes to fulmars to eiders that breed here too. 'Afraid I don't come with subtitles,' he adds, referring to his broad Geordie accent – he moved here in 2015 after more than a decade of wardening on Northumberland's Farne Islands – 'but you can ask me anything you want, even the football scores!'

Asking, in this season of a quarter of a million nesting seabirds, about the grey seal breeding colony that establishes itself here each autumn seems almost as perverse as his suggestion to quiz him on football. Nevertheless, as the last passenger staggers up the slope from the jetty, lugging a huge cool box with 'Keep Cool and Carry' on the side while ducking the dive-bombing beak of a territorial Arctic tern, I hang back for a quick seal chat. This is the first of the east coast pupping locations I've visited where births have surged in recent decades and, as I travel on in the months ahead, I'm hoping to learn more about this phenomenon and determine if it's as positive as it sounds. I've read that numbers initially grew here from the second half of the 1970s, due to the displacement of breeding females from the Farnes where an intensive cull had been underway, but that doesn't account for the continued increase. Around five hundred pups were recorded in 1980 – now, the total is five times that figure.

Though David has other visitor engagement duties to perform, he gives me a brief insight into life here during the pupping months. 'We close the island to the public at the end of September so the seals won't be disturbed,' he says, indicating that only Scottish Natural Heritage personnel and research scientists are permitted to stay at that time. Unlike Skomer and Ramsey, where mothers and pups are confined to beaches at the base of high cliffs which can't be reached by members of the public, parts of the Isle of May are low-lying, meaning that the ever-expanding numbers of seals are able to move up from its rocky margins and colonise the interior. 'You can't access this jetty either when the cows and pups spread. I have to inform the Port Authority and say this part of the island is closed to all boats. The seals slowly but surely take over till you've got them literally right outside your front door.'

Where the Seals Sing

As far as the scientific studies go, I understand, from my preliminary reading, that a wide range of different findings, from
behavioural to microbial, is being gleaned from the Isle of May
breeding population. In addition to the presence of SMRU scientists, researchers from Durham University have explored such areas
as cows' contrasting mothering styles, their varying responses to
stressful stimuli and how this impacts on reproductive success.
Over five successive breeding seasons, several cows have been
fitted with a specially modified heart monitor, cables and electrodes
that transmit information about their heart rate patterns to a nearby
receiver. Essential knowledge about their physiological response
to, and recovery from, stressful events such as aggression from a
neighbouring cow or unwanted attention from a bull has been
acquired. It's been discovered that though behavioural observations
may suggest that a cow's returned to a relaxed state quite quickly,
her heart rate may remain raised for some time. In the future, in
the context of, for example, wildlife tourism, some committee will
no doubt be deployed to determine what constitutes an 'accept-
able' level of anthropogenic disturbance of seals, and stress
response findings of this kind could prove to be crucial.

When another tour boat, a rigid inflatable this time, bimbles into
the harbour and moors behind the *May Princess*, David apologises
for having to curtail our conversation and suggests that we have
a longer phone chat at some point. I leave him to welcome this
latest batch of visitors and head uphill to access one of the marked
footpaths. As the island's not used as a summer haul-out, no more
than a handful of seals are likely to be lingering around its fringes.
But since it's only about a mile long and I have three full hours
ashore, I'll hopefully have time to locate a few.

It turns out to be impossible, though, to focus on seeking seals
when the seethe of seabirds is so overwhelming. Each species is
at a different stage in its breeding season: while the fulmar chicks
are floofy white hatchlings with tiny tubular nostrils, a few of the
young Arctic terns are advanced enough to take to the wing.
Meanwhile, as was the case above the jetty, adult Arctic terns are

swooping and screaming to defend their nest sites, or coming in to land with glints of sandeels in their beaks.

I'm unexpectedly drawn to the shags, the adults with their emerald eyes and the chicks still downy grey rather than lustrous black. Already almost as big as their parents, some are standing, beaks open, and motionless but for a flutter of their throat muscles, a form of panting to dissipate the heat. Others are flapping and flexing their wings in anticipation of their first flight. A parent riffles its bill through its sleepy offspring's feathers, preening it clean of parasites. Another chick makes a lunge for the regurgitated fish in its parent's throat, thrusting its whole beak and face into the yellow gape. Nest remnants are visible too – not just sticks and seaweed but a flip flop, plastic bottle tops, half a biro, a piece of rope.

Throughout the afternoon, the lighthouse, with its central, elevated position, supervises my wanderings and its castellated tower looks just as incongruous in close-up as it did from the sea. Designed in the nineteenth century by Scotland's premier lighthouse engineer, Robert Stevenson, grandfather of Robert Louis, it was referred to as his 'toy fort'. It's linked to other buildings – the old engine room, the keepers' cottages in which David Steel and the seasonal scientists live – by named paths. The steep Shag Brae has nothing to do with the bird for which I've developed a sudden affection: also known as Palpitation Brae, it was apparently coined by naval personnel stationed here in the Second World War who were 'shagged out' after climbing it.

I walk on towards the lower-lying, north end of the island, all the while trying to picture it thronging with seals in autumn, the soundscape not the stramash of seabirds but howling cows and yowling pups. A smaller copybook lighthouse, which now houses the Isle of May Bird Observatory, the second in the UK to be established after Ronald Lockley's on Skokholm, is sited here. On an adjacent rotary washing line, the laundry of its volunteer staff is drying in the sun.

I've reached the rocky peninsula of Rona – almost a separate island – which, as a research area, is out of bounds. Since its name possibly stems from the Gaelic *ròn*, meaning seal, it pleases me to spy, through binoculars, seven cows recumbent on the rocks. I decide to settle here for the remainder of my shore time, enjoying the warmth and the view of the slumbering seals. Their languid movements are such a contrast to the frenetic bird activity – even the head and belly scratching into which they usually put exceptional effort lacks intent today. Thanks to fur and up to ten centimetres of blubber, they're at risk of overheating when hauled out on days like this so they periodically raise and waft a front flipper or fully fan out those at the rear for maximum heat dissipation on exposure to the air. One of the effects of their elaborate vascular networks, known as *retia mirabilia* or 'miraculous nets', is enabling blood to rapidly reach the surface of their skin, including that of their blubber-free flippers, to help promote cooling.

It's so rare to have a windless, scorching seal-watching experience that, like the cows, I soon find myself sinking into sleep, wishing, as I do so, that Dad's vascular networks were as miraculous as the seals', with blood freely ferrying oxygen to his brain through vessels that would never narrow.

I also drowsily recognise how different my brief experience of the Isle of May has been from Robert Louis Stevenson's introduction to the area. At the age of eighteen, he travelled to Anstruther and quickly formed an irreparable impression.

'I am utterly sick,' he wrote in a letter to his mother, 'of this grey, grim, sea-beaten hole.'

Four hours later and there's a grey, grim, sea-beaten throb behind my right eyeball. Too much sun. Too much sun dazzle. Migraine.

Having boated back to Anstruther, I've travelled the short distance to Pittenweem, the next fishing village along the coast, where I've booked accommodation for the night. All I want to do is retreat to my room, shut the curtains against the still-blaring sun and lay my head on a cool pillow. Instead, though, I'm sitting on the sea wall to get the strong phone signal I need to make my evening call to Dad.

After reminding him of my location as usual, I hear about what he's been doing. Wednesday, isn't it? Ah yes, Tuesday. Lovely weather, walk round the lake, cygnets getting bigger all the time. Lunch – can't remember. Yes, whatshername, the carer, came. Puzzles in the paper. May have had a nap. *Pointless. Eggheads.* Supper – can't remember – oh, wait a minute, fishcakes and potatoes and some green stuff. No, not broccoli. The other one.

He sounds fine. 'What about mail? Have you had any letters that I need to see to?'

'Haven't we already talked about that?'

'Not today. Not for a few days, actually.'

I've devised a system whereby he files all his mail in a purple folder – when I'm with him, we go through it together and when I'm travelling, he reads out whatever seems pressing and I do my best to keep on top of it from afar.

'Mail. Right. Now where's that going to be?'

'In your purple folder, hopefully. On the table next to the phone.'

'Bugger. It's not there.'

With my free hand, I massage my right temple and brow. Relax. It's a given that he's going to fumble and faff. I'm facing north, away from the sun, away from the sea, towards a terraced row of cottages. One has a mini-rowing boat full of geraniums attached to the wall. Outside another, there's a rusty bike with lobelia spilling from its front and back baskets and scallop shells tucked between the spokes. A third has a name plaque, Seal Point, and I find it irritating in the extreme that the accompanying image is a sea lion, not a seal at all.

'Got it. In the kitchen. Next to the toaster.'

'Great.'

I hear the rustle as he pulls his mail from the folder and starts shuffling through it. 'Red Cross – they want more money out of me. Bird magazine from the wotsit – the RSPB. Phone bill. . .'

'No need to worry about that. You pay by direct debit.'

'Summer raffle – Guide Dogs for the Blind. Some bloke telling me I ought to get a smart meter. Arthritis Care or whatever they're called – they're still writing to your mother. And something about insurance.'

'What insurance?'

'The one I renewed when they phoned me.'

An Arctic tern's stabbing the place behind my eye with its beak. I hear more rustling as Dad starts to shove the mail back into the purple folder.

'Dad, wait a sec. Listen to me. What did you renew? Car insurance, house – they're all up to date. Who asked you to renew something?'

Silence. Too many questions at once. He'll be getting flustered.

'What does it say in the letter?'

'That I'm covered for another year.'

'What's covered?'

'I'll tell you now – stop making such a hoo-hah.' There's some unintelligible muttering as he reads the letter aloud to himself. 'Look, it says right here. Cooker. Dishwasher. Freezer. Fridge.'

The door to one of the cottages opens and a German shepherd bounds out, followed by a woman in a raffia sunhat.

'Dad!'

'What?'

'You've never insured any of those things before! And you haven't got a bloody dishwasher!'

The woman glances towards me from beneath her broad brim, then quickly looks away again. I swing my legs over the sea wall so that I'm facing away from the cottages and continue more quietly.

'You didn't give your debit card details over the phone, did you?'

'No! Yes. I might have done. If they asked for it.'

'Can't you remember?' Tactless question. 'Listen, Dad, it's a scam.'

'A what?'

'Someone phoned you and sold you fake insurance and took your debit card information. And now they can use it again – they've probably been using it all week.'

'Of course they haven't! I'm still using it myself!'

Throb. Throb. Throb. My eyes seek out an alternative to sea, find a blessedly dark mass of rock in the middle distance. Something like this almost occurred once before – early last year, I think, a while before his dementia diagnosis. I luckily happened to be visiting when he answered the phone and was able to intervene just as he was about to read out his card number. The caller rang off as soon as I started asking questions.

'Look, Sue, how can it be a whatchamacallit – a scam? It says here at the top of the letter: "For Your Peace of Mind—"'

'Anyone can pretend to offer that!' My patience, ordinarily a stable cliff, has been eroded to a wonky sea stack. 'I've told you so many times never to give your details out over the phone. Told you to screen your calls and put the phone down if anyone tries to sell you something. Why did you do it? It's going to be a nightmare trying to sort everything out from up here.'

There's a long silence – during which I'm startled by the sight of a bull seal labouring onto the rock I'm watching – followed by a muffled gulp.

'I'm no good to you. No good to anyone.'

Not a gulp, a sob.

'Don't say that. Of course you are.'

'No, I'm bloody useless.'

'Dad, I'm really sorry. I didn't mean to shout. I'll make sure everything gets sorted. Okay?'

The bull shunts his bulk around to find a comfy spot and I likewise try to shunt our conversation onto more comfortable terrain. 'I saw puffins today. And kittiwakes. They're much smaller than the herring gulls you see down at the lake. And very noisy.'

Bit by bit, bird talk eclipses scam chat and we eventually get to the point where Dad's calm and settled enough for me to ring off.

And, like a bull seal lumbering from the water, I haul my worry about his money, my guilt at having upset him, and the kittiwake shriek of my head into the B&B and up the stairs to bed.

I wake with residual pain and the feeling that my frustration with Dad has found its proper target, mutating into anger at those who dupe the vulnerable and more advanced in age. This rage intensifies when I check out the supposed insurance company online. At first glance, their website looks kosher, with images of suits shaking hands with clients whose smiles proclaim their Peace of Mind. But deeper delving unearths lots of unattributed testimonials and a parent company, purportedly authorised by the Financial Conduct Authority, of which I can find no trace. I'm able to access Dad's bank account online too and find debits of £149 for each 'insured' household appliance plus five other dodgy payments for amounts ranging from £52 to £190. I obviously have to cancel his card but I also need to make sure he's got enough cash to cover the time until the new one arrives.

In addition to trying to sort out these financial complications, I have two seal-related video calls scheduled for today. As a follow-up to Ailsa Hall's description of her research and my preliminary reading on the various scientific projects that are underway on the Isle of May, I'm going to be speaking with a couple of other academics who've been undertaking a major, long-term physiological study there.

Dr Kimberley Bennett of Abertay University leads the PHATS team, which stands for Pollutants, Hormones and Adipose Tissue in Seals. In both her official online staff photo and more casual snaps in which she's wearing full field research gear, she manages to look glossy and groomed, in acute contrast to my current strained, post-migraine face. When we meet on screen, though, her sleek hair becomes rumpled and her cheeks get progressively pinker as she shares her huge enthusiasm for her work.

'Seals can do such extreme things as routine even when they're babies, things we can't even contemplate,' she says. 'And I love understanding, from a physiological point of view, how they deal with different stressors and challenges.' One of the biggest anthropogenic burdens they face is environmental contaminants, and she and her team, which includes Ailsa Hall, are researching the impact of persistent organic pollutants, or POPs, on grey seals' fat tissue function. This group of pollutants includes the insecticide DDT – the indiscriminate use of which inspired Rachel Carson's seminal text of 1962, *Silent Spring* – and polychlorinated biphenyls (PCBs), that, due to their resistance to fire, used to be omnipresent in electrical appliances, building materials, paints, plastics and packaging. When POPs were discovered to be carcinogenic and endocrine disruptive, suppressing immune function and causing reproductive abnormalities in both humans and wildlife, bans started to be introduced in the United States and Europe in the 1970s and 80s. The Stockholm Convention, a global treaty adopted in 2001, requiring its signatories – initially ninety-two nations plus the EU – to eliminate or reduce the production, use and release of POPs, was another major milestone. Although, following these various international actions, toxicity levels in the wider environment were at first seen to fall and some wildlife populations seemed to recover, marine mammals, as apex predators, were still found to be at great risk as the pollutants bioaccumulate and exist in the highest quantities at the top of the food chain.

A recent research paper authored by PHATS scientists confirms that POPs are even now continuing to linger in the marine environment and that the problem is far from resolved. POP levels in blubber samples collected by Kimberley and co. on the Isle of May were compared with the levels that were recorded in pups there fifteen years previously and though the amounts of some toxins had decreased, other contaminants, including DDT, had fallen very little during that period. More alarmingly, experts at the specialist laboratory at the University of Liège in Belgium, who assisted with the preparation and analysis of the blubber samples from the first year of the project, maintained that the pups, who had not yet been to sea or eaten any fish, were still manifesting some of the highest concentrations of POPs they had ever detected in an animal. Pollutants that had accumulated over the years in cow seals' blubber through eating contaminated fish were being transferred into the milk on which they fed their pups. The fact that pollutants adhere to, and become concentrated in, the microplastics that seals unwittingly ingest must also be adding to the burden.

The PHATS team have further discovered that there are metabolic implications, as the higher the concentration of PCBs that a seal pup was found to have in its blubber, the less glucose was used by its fat cells. While the consequences of this impairment in blubber function are not yet fully understood, pups' capacity to reach a healthy weaning weight, and thereafter their chances of surviving their first year, are conceivably being seriously compromised.

Having filled me in on some of the outcomes of the PHATS project thus far, Kimberley sinks more deeply into her cosy home sofa and describes the process of obtaining the blubber samples during each seasonal six-week stint on the Isle of May.

'We usually work with thirty to forty animals each year, anaesthetising and taking fat samples from the pups once when they're suckling and again when they're fasting after they've been weaned,' she explains. 'One of my jobs is to go out with the catching team and have the flask ready for the blubber to take back to the tissue culture lab we've set up. I'm the runner, I have to get in and out of the colony each time a sample's taken, so I wait there planning the best route, the one that'll cause minimal disturbance. Some females roar at me and don't do anything while some want to eat me from two hundred metres. That's always a bit of an adrenaline rush, getting through in a way that's safe for me and the animals. As you know, grey seals are very wary of people – on land, they're really skittish and can be very aggressive, and rightly so. I've worked with elephant seals on South Georgia in the moulting season and they didn't care about people at all. It's very different working with an animal that just opens one eye and snorts at you from an animal that's prepared to attack your kneecaps!'

Setting up the temporary lab on site also presents challenges. 'We've got to make sure we keep the tissue as clean as possible – we have to set up a little area of cleanliness in a place that's used in the summer for dealing with bird scats, owl pellets, dead birds. And we're constantly having to deal with mice in the lab.'

I've read about the Isle of May's unique tribe of mice as well as the claim that their population's as many as six thousand. In the 1980s, a scutter of house mice was released on the island from Orkney, an experiment in gene hybridisation with the mice that were already resident. They interbred, multiplied and appear to be a slightly larger version of those that inhabit the mainland.

'My PhD student a few years ago left blood smears from the pups out to dry before staining them up and the mice licked all the blood off,' Kimberley continues. 'And I'm always worried that our cables will get chewed.' They have a habit of scooting over sleeping scientists in the middle of the night too.

Along with the flourishing mouse colony, she's also witnessed the grey seal colony's growth. 'Since 2000 when I was doing my PhD, it's changed so much – our range has been dramatically restricted in terms of what we can do in our spare time as the seals have moved to new sites inland. I think the colony's been through its massive expansion phase now, though, and won't increase very much in the years ahead.'

This chat with Kimberley has given me much on which to ponder, not least because I'm now seeing the rise in births on the Isle of May in the context of the threat that the contaminants burden arguably poses to pups' post-weaning survival. I don't immediately have time to process all I've heard, though, as there's a prolonged and convoluted conversation with the fraud department of Dad's bank to contend with before video call number two gets underway.

Post-doctoral research fellow Kelly Robinson is part of Kimberley's PHATS team. She speaks to me from her office, an unexceptional space but for the two cuddly seals on the shelf above her head. In all her work photos online, she's cocooned in waterproofs, gloves and hat but in their place today are a flowery top, multiple finger rings and a long auburn plait.

Her PHATS project role includes assisting her fellow scientists to obtain the blubber samples from the Isle of May seals each breeding season, then basing herself for several months at the laboratory at the University of Liège to prepare the samples for analysis. She endorses Kimberley's view of the difficulty of maintaining clean conditions in their workspace on the island when surrounded by 'cannibal mice' and speaks, too, of the contrast between engaging with the life force of their study animals in the 'absolute carnage' of the colony and working with small blubber extractions in the sterility of the university lab.

'I love that I've been able to collect pretty much all the samples that I've ever worked on myself,' she says, 'so that months later, when I'm looking at a sample in the lab, I can instantly visualise that individual. I love that I can think oh, I know why this label's smudged – it's because I fell over when I was collecting it. It could otherwise be very abstract – a tube with some liquid in it. It's really important not to take the samples for granted. The work requires you to handle the animals and that's controversial in some instances – it's why we have ethics committees and Home Office regulations to make sure that if we do something it's for a very good reason – so I personally think it's great to have an appreciation for not just the samples but the seals themselves.'

'Does appreciation for the seals ever evolve into attachment?' I ask, assuming my question will evoke disapproval and she'll emphasise the need for scientific objectivity.

'I've yet to come across anyone in all the teams I've worked with, however hard the exterior they try to project, who isn't emotionally attached. If a natural thing, something that's not caused by humans, happens to one of our study animals, we're not meant to intervene and it's really sad to sit by and watch terrible things happen to animals you've grown attached to. If you don't care about a study animal you won't treat it with the respect it deserves. I always really hope I'll see some of my pups again even though first-year survivorship is absolutely abysmal.'

This acknowledgement of pups' struggles prompts us to talk about the high concentrations of POPs detected in the blubber samples on which she's been working. She tells me that contaminants that were manufactured before the bans were introduced are still making their way into the environment whenever the materials and buildings that contain them get discarded or destroyed. PCBs that are already in circulation are long-enduring too: originally favoured because of their extreme heat-resistant properties, they're exceedingly robust and stable.

However, in spite of affirming that the toxic burden will continue to impact on seals' health in the years ahead, she hopes that levels will gradually fall and, in the meantime, suspects that greys, as a population, will be robust and stable enough to withstand it. 'You've only got to look at some of the wounds they have to deal with – they just keep on going – and their natural curiosity, despite the potential dangers around them, never ceases to amaze me.' For the remainder of our conversation, her objectivity goes into retreat again as she fully reveals her zeal for seals. 'They're such personalities – that's anthropomorphising hugely but it's now scientifically proven that they have what are known as distinct behavioural syndromes. I love how much attitude they have and how temperamental they can be, going from happy to curious to absolutely rage-filled in the space of a few seconds.

'I just love that you never know what you're going to get.'

Over the next few days, out of all the information and observations that the PHATS scientists offered, one rueful comment by Kimberley, in particular, keeps coming back to me. 'The more I work on this project, the more I think we're measuring the mess that's been created and quantifying the mess.'

As far as quantifying the mess surrounding Dad's scam incident is concerned, the bank's reviewing whether or not to refund the money he's lost and I've signed him up to a call protect service which should filter out future unwelcome approaches. Bethan the carer has stuck a notice on the wall above his phone, warning him against scam callers, and whenever I speak to him, I issue a strong reminder about this too. 'Well, of course I wouldn't be daft enough to give out private information to some wally on the phone,' is his usual response.

While I at first feared I'd have to dash back to Wales again, the situation feels sufficiently sorted for me to embark on the next phase of my journey, south to Northumberland. On the way, I decide to make a final Scottish stop for tea and cake in Eyemouth, about five miles north of the border. On the road leading down to the town, pennants hang from all the street lamps, featuring photos of Eyemouth Herring Queens from years past. 1953 – Miss Katherine Dougal. 1958 – Miss Terry Chisholm. 1971 – Miss Christine Hastie. Those from the more distant decades are decked in long dresses and clutching an orb.

In the narrow harbour, there's a bevy of berthed boats, some named after women (*Maxine*, *Sally-Ann*), some with gung-ho intent (*Crusader*) and some (*Good Fortune*, *Fruitful*) in hope of successful fishing. The fact that both the Contented Sole pub and the take-away fish stall on the quay are bustling with customers suggests that today, at least, Good Fortune has prevailed.

As I draw closer to the kiosk, however, I find that its customers aren't feeding themselves at all but are heeding a series of notices urging them to Feed The Seals. I elbow my way through to the harbour railings and lean over. Floating on the water below is an orange platform onto which a large grey seal bull has hoisted himself. Above his head, a fish is swinging, attached by a clip to a line and pole, held by a man in flip flops on the quay. Whenever the bull stretches up to grab the fish, the man jerks it just out of reach to the accompaniment of bursts of laughter from his mates.

About ten metres away, a cow is swimming back and forth, keeping her distance for now, while closer to the harbourside, a second cow is maintaining a vertical position in the water, motionless except for the vigorous flaring of her nostrils. Her eyes are pink-rimmed and failing to focus and I realise she's putting extra effort into scenting because she's blind.

An American child who's obviously been to too many performing sea lion shows complains that the bull won't leap up to snatch the fish or do any other tricks.

'Where do they live when they're in the water?' he asks his father between whinges.

'They live in the river, kiddo. They live on a rock.'

For the next hour, I watch umpteen tourists take their turn to dangle fish over the railings. Though the bull monopolises the feeding, one woman at least takes pity on the blind cow. She asks me to hold the lead of Hugo, her golden retriever, while she buys a cupful of fish. £2 buys four pieces, £4 buys nine. She spends £8 and feeds them all to the blind cow, well away from the bull on the mini-pontoon.

This is the closest I've been to a seal since last year's visit to the RSPCA rescue centre in Taunton but I'm feeling more discomfort than delight. Eyemouth is clearly a busy harbour and any seal that frequents it risks serious injury from boat propeller collisions, as has occurred in a number of Cornish resorts. Diesel poisoning is another hazard that's suspected to have caused the deaths, by slow organ failure, of two Cornish seals. Tempting though it may be to try to connect with wild animals, feeding builds dependency too, leading them to become reliant on the handouts and less efficient at hunting and foraging for themselves. It's also been established that when an animal loses its natural fear of humans and associates them with the provision of food, it may become increasingly aggressive if the expectation of being fed isn't met. Equally worrying is the fact that seals who become habituated are likely to be an easier target for those less enamoured of them, as was the case when R. H. Pearson's Diana was killed.

At 4.55 p.m. precisely, the guy manning the kiosk emerges, wearing his white fishmonger overalls, and starts putting the poles and lines away, prompting the throng of people to disperse. I wander over to speak with him and, though he's quite taciturn, he tells me that Feed The Seals was set up over a decade ago and that every morning, he arrives to find the bull – 'the boss' – and the two cows in the harbour waiting for him.

After he drives away, none of the seals is inclined to leave, vying with hooligan juvenile herring gulls for leftover scraps of fish. I'm about to go in search of the cake I came for when a couple of adolescents canoodle their way to the railings. They pause in front of the Feed The Seals sign that's been left on display, then stare down into the water.

'Shit, Kate, what are we s'posed to feed them?' the boy asks, construing the sign as a command rather than an invitation.

'I got a Mars Bar in my bag.'

'You can't give them that! They'll go belly up.'

16

The Hubris of Scuba

Like an exuberant Labrador, the young seal bunts the upper arm of the diver who responds by stroking her head with a hand clad in neoprene gloves. His other hand somehow manages to keep operating the underwater camera as the seal nuzzles up against his chest then glides higher, first to brush his mask with her whiskers, then to tug on his wetsuit hood. Through the diver's bubbled exhalations, the seal flippers and mouths his rubbery head-covering before finally appearing to settle on his shoulders, a graceful, weightless version of a fireman's lift.

The diver featuring in this footage is Ben Burville, a Northumberland-based GP who devotes as many non-work hours as possible to exploring the waters around the Farne Islands, interacting with grey seals and sharing the visuals online. He's been uploading videos to his YouTube channel since 2006 and has an active Twitter presence too, with liberal use of the hashtag 'divebuddy' in all his seal-themed posts. I spend my first evening in Northumberland binge-watching his films, witnessing a succession of seals approach him in the spirit of play and curiosity, permitting him to take their front flippers in his hand and revelling in having their bellies scratched.

He claims to have spent more time underwater with grey seals than anyone else on earth and has captured behaviour, including non-vocal communication such as foreflipper clapping by bulls, that's never been seen before. While I recognise that he's contributing to the expansion of seal science and his sequences of seals' subaquatic ease, as compared with their on-land gait, are undeniably appealing, I can't help but feel troubled by what I'm seeing too. The tabloid press periodically seems to latch onto his grey seal encounters and his footage appears in viral video compilations that get circulated ever more widely. Headlines like 'Lovable Seal Hugs Diver', 'Wild Grey Seal Gives Doctor Loving Embrace' and 'Diver Makes Friends With A Seal Who Smothers Him In Kisses'[1] accompany the stories, giving the impression that Burville's engaging with cute, cuddlesome, semi-domesticated creatures. Yet they're wild animals whose mouths contain lethal teeth and hazardous bacteria and who have the capacity to predate on other large marine mammals such as the harbour porpoise, as evidence from the coasts of France, Belgium and Holland, as well as Ramsey Sound, has shown. Those who might feel encouraged to try to emulate his interactions, whether at sea or on land, risk causing not only wildlife disturbance incidents but also injury to themselves.

High above the North Northumberland harbour of Seahouses, peak season visitors make fractious circuits of the car parks, seeking an empty space. The lucky ones disgorge kids and pushchairs, dogs and windbreaks, cool boxes and bodyboards, buckets and spades. Some flounder down to the dune-backed beach while others head for the kiosks lining the harbour to queue for tickets for a boat trip a few miles offshore to the Farnes. Later today, I'm meeting Ben Burville to find out more about his seal-diving experiences but first, I'm going to join one of the boat trip queues in the hope of seeing some seals for myself. Like the Isle of May, the Farne Islands support a large and flourishing autumn pupping colony but, unlike the May, the archipelago remains a busy haul-out all year round.

Bypassing the membership van for the National Trust, who own and manage the islands, and all the ads for seabird and sunset trips, I eventually manage to get myself booked onto a grey seal cruise that's leaving within the next half hour. The wait, more of a jostle than an orderly line, is accompanied by a boy in a FUN FUN FUN T-shirt scanning the faces of passengers disembarking from completed trips and asking 'Who's been sick?' Only the young couple to my left seem unbothered by the scrum – the woman travelling light with just a banana curving from the pocket of her shorts and the man in a sunhat with a neck flap sensually stroking her back.

Once we board, the smell of fried bacon from a food booth mingles with diesel and bilge. A female eider and two large duck-lings flap-and-paddle out of our way as we depart. The grey of the early morning sky has cleared to white clouds balled like bog cotton while a sprightly breeze already has the woman next to me complaining that she's left her cardi behind.

After urging us to stay seated as no side of the boat's 'a bad side when it comes to seeing seals', the commentary gets into full flow. The Farnes archipelago comprises twenty-eight islands at low tide but only around half that number at high. It's split into two island groups: Inner and Outer, separated by Staple Sound. If we look north along the Northumberland coast we'll see majestic Bamburgh Castle, while the ruins of Dunstanburgh Castle teeter on an outcrop to the south. At this time of year, the Farnes host nationally and internationally valued seabird colonies with over eighty thousand nesting pairs: species include kittiwakes, guillemots, Sandwich and Arctic terns, eiders, puffins, cormorants and shags.

When we pass Inner Farne, the closest island to the mainland, squalls of Arctic terns, seeking to protect and feed their young, are visible through binoculars while visitors whose chosen cruise includes an hour ashore process along the cliffs, using a variety of defences from wide-brimmed hats to umbrellas against the birds' dives.

'I've seen a seal! I've seen a seal! I've seen a seal!' the FUN FUN FUN boy shouts over the top of the commentary as we continue our journey along the edge of the Inner group.

There are, it turns out, about forty seals – cows and bulls, adults and juveniles – piled on low-tide rocks and mats of kelp. Considering they're so densely packed, I'm surprised there's no grouchy howling and instead of lunging and shunting to maintain their personal space, they're enjoying deep repose.

The woman who's forgotten her cardi seems strangely disapproving of their inactivity. 'Well, those seals don't look like they're going to do very much else today. They're not going to do their ten thousand steps.'

Most passengers are on their feet, clustered along the left side of the boat, flourishing their phones. Kids are leaning over the side to wave hello. A toddler at the stern's on the whimpering brink of a tantrum because she can't see anything. The commentary about grey seals' diet to which no one's listening turns into an announcement by the skipper that we're going to move in 'quietly' for a closer view.

'I've seen a seal! I've seen a seal! I've seen a seal!'

Inevitably, the combination of boat and brouhaha rouses the seals and hustles all but a handful from rocks into sea. The commentary plays it down by saying they're a nervy bunch and adds that it's unusual for them to be hauled out around the Inner Group of islands. If we head on now to the Outer Group, it's likely that we'll see many more.

Everyone settles back down into their seats for the crossing of Staple Sound and focuses on scrolling through their photos while I inwardly fret that the seals' essential rest was so utterly disrupted. Wildlife tourism's become a lucrative industry in the UK and seals are regarded as a major asset thanks to their fidelity to, and predictable presence at, known haul-outs. Unlike trips to watch other marine mammals such as dolphins or whales, sightings can be pretty much guaranteed. Tide permitting, seals may also haul out for hours at a time, which further increases their worth to boat operators as there's the potential for scheduling a whole series of cruises. While some parts of the UK have codes of conduct in place – the Pembrokeshire Marine Code, for example, and the Cornwall Marine and Coastal Code – which encourage members of the public to behave responsibly when in or on the water, and a national training programme, the WiSE scheme, enables boat operators to gain 'wild-life-safe' accreditation, compliance is entirely voluntary. The volume of visitors and commercial operators here at the Farnes is enormous compared with other seal sites to which I've taken a boat and my fear is that the disturbance impact I've just witnessed may recur multiple times in the course of a single day.

The commentary switches from wildlife facts to Victorian melo-drama as we approach Longstone Island and its lighthouse, red and white striped like a giant traffic cone. From here, in 1838, the daughter–father team of Grace Darling and lighthouse keeper William saved several shipwrecked survivors from the rocks of Big Harcar, another island in the Outer Group, by rowing a coble, a small, open boat, across stormy seas to reach them. The Victorian public was enthralled by Grace's role in the rescue and, thanks to her youth, gender, courage and perceived virtue, she was elevated to the status of the nation's heroine. Portrait painters made the journey to her home to capture her likeness and she received a gold medal for bravery from the Royal Humane Society as well as numerous gifts, letters and marriage proposals. Her death in her father's arms from TB at the age of twenty-six only served to inspire further glorification.

Gatherings of seals around the Outer Islands turn out to be smaller than expected – perhaps, says the skipper, because 'there's been a lot of diving activity here in the past few days'. There's again a mix of adult bulls and cows plus juveniles, some resting on boulders and others on exposed forests of kelp, its abundance a vivid contrast to the paucity of terrestrial vegetation. Having seen the earlier seals, my fellow passengers' reaction to the new sight-ings is much less feverish. Most stay seated and the FUN FUN FUN boy launches into an 'I need to pee' chant, prompting his mother to hand him a packet of Wotsits to distract him. The young couple, meanwhile, are still sequestered in their love bubble, applying sunscreen to each other's knees and thighs.

The names of several of the islands we're skirting

– seabird-stacked Staple, the Brownsman, North and South Wamses – are familiar to me from my recent phone conversation with David Steel, the reserve manager on the Isle of May who used to be head warden here on the Farnes. Because he was busy with visitor arrivals when we met on the May, he suggested I contact him at some stage for a less pressured chat. As soon as I did so, he started speaking of his years here, especially each pupping season, with immense enthusiasm. 'I must apologise, Susan – if I talk too much, just tell me!'

He explained that he lived in a cottage on the Brownsman and though no pups were being born on the island in 2001 when he first started wardening, by the time he left almost a decade and a half later, Brownsman births were totalling over seven hundred. 'It was an amazing experience to live among the seals. Superb. In November, it got dark by four p.m. – after that, I couldn't go outside because of marauding bull seals. It was a bit like *Jurassic Park* – you could hear the bulls fighting and fending each other off outside your door. One cow gave birth right outside too – I'd often find her pup inside the house if I forgot to latch the door properly. A few times, I found it in the bath!'

Like the Brownsman, Staple's now 'a major player' as far as pup numbers are concerned, with cows observed to be shifting from the lower-lying Outer Islands where there's greater pup mortality and colonising those with higher ground to gain sanctuary from storms. In recent times, pup nurseries have also become established on some islands of the Inner Group. Across the entire Farnes archipelago, well over 2,500 births are now recorded annually, an increase of more than 50 per cent in just five years.

This latter-day boom, though, tells but a fraction of the story of grey seals' presence on and around the islands. There's documented evidence of their existence here from over six hundred years ago when a small cell of Benedictine monks based on Inner Farne traded with their parent house of Durham, providing seals to be used for food and oil: a receipt from 1371–2 details that '6 celys' garnered them '27s 4d', over £700 in today's money. Larger-scale and more organised exploitation continued in subsequent centuries until the Victorian era, when the Inner Group was leased, then purchased, by Charles Thorp, Archdeacon of Durham, whose progressive attitude to conservation temporarily afforded both seals and seabirds some measure of protection.

In the twentieth century, the Farnes seal population became one of the most thoroughly studied but also the most persistently persecuted in the UK. In response to complaints that excessive preservation of seals had upset the balance of nature and was causing severe damage to the salmon fishing industry, a programme of marking, counting and weighing pups was implemented in order that an informed assessment of the size and health of the colony could be made. In spite of the fact that Berwick Salmon Fisheries recorded their most successful year in 1957, and researchers felt that the colony should remain untouched until further information had been gleaned, an announcement was made in the House of Commons in 1958 that a cull would take place. However, having been described in a newspaper headline as 'Murder in the Nursery', and by Farnes naturalist Grace Hickling as 'a most distressing experience' because the pups 'proved unexpectedly difficult to kill; the details are too harrowing to relate', it was eventually suspended.

Further culls were initiated in the 1960s and 70s, at first on account of fisheries conflict again and later because breeding seals were perceived to be eroding vegetation and damaging puffin burrows. 'To think that that was believed to be a justification to kill them!' said David Steel. 'You think *nah, they'll never get away with that* – it's barbaric.' Over 1,500 cows and 1,600 pups were shot between 1972 and 1977 and other adult females are believed to have relocated to the Isle of May. The current bumper numbers, therefore, need to be contextualised – pup production on the Farnes was recorded as 2,041 in 1971 so it could be argued that more recent totals represent a rebound and recovery to that pre-intensive-cull figure rather than an unprecedented expansion.

I asked David what method of pup monitoring was favoured during his wardening years here. 'We'd go out every three to four days and colour-mark them,' he said. 'And we'd change the colour each time, using four in rotation, to keep track of the newborns. Even after twenty years, I get caught out trying to age a pup – spraying makes the job much easier.' He acknowledged its disturbance potential, however, and added that in the seasons since he left the Farnes, pup counting via drone technology has been trialled, with the pilot flying the drone at a height that causes no distress to the seals whatsoever.

He referred to another example of pioneering grey seal research that originated on the Farnes too. Back in 1951, stainless steel tags engraved with a serial number and contact address were attached to the hind flippers of ten pups in the hope of gaining an insight into their movements post-weaning. 'Believe it or not, three weeks later a letter arrived – seal number one was found alive on a beach near Stavanger!' At the age of just six weeks, it had swum four hundred miles across the North Sea to Norway. By the end of the decade, nearly 2,500 pups had been tagged on the Farnes with recoveries all along the east coast of the UK, elsewhere in Norway and even on the Faroe Islands over five hundred miles away.

'Don't eat it if it's been on the floor!' The mother of FUN FUN FUN, whose voice is just as strident as her son's, plucks a grubby Wotsit from his hand and chucks it overboard. Our grey seal cruise is approaching its scheduled end and we're steaming back to Seahouses into the breeze, the young couple snuggling under a single fleece and my neighbour reminding us she still craves her cardi.

As we join the queue of returning boats waiting for a berth at the harbour – *Serenity 2* and *Glad Tidings VI* ahead of us, *Ocean Explorer* slotting in behind – the skipper feels the need to apologise for the fact that few seals showed up around the Outer Islands, prompting me to recall David's words about the evolution of the tours.

'When I started working on the Farnes in 2001, there was a well-known fisherman – he had three boats – a big grizzly bear of a man – and I ran into him in The Ship, a Seahouses pub. I was an impressionable twenty-one-year-old at the time and I was terrified, 'cause here was this guy saying – pardon my language, Susan – "I hope you're going out there and shooting those bastard seals." And, no word of a lie, by the time I left in 2015 this same bloke would be meeting me in the pub and going, "How are the grey seals? Are they doing okay? I saw a pup and I was a bit concerned about it." There was a big, noticeable shift in attitude and he wasn't the only one. The difference in the fourteen years was that seals were bringing in a livelihood for them via the tour boats.'

The volume of wildlife-watching trips here may well be causing the seals undue stress and there's plainly the need, at the very least, for additional education and awareness campaigns for both the commercial operators and their customers. Yet it's sobering to realise that, in the context of their historical victimisation, Farnes seals are currently leading comparatively tranquil lives.

Ben Burville's running late. We have a narrow window for our chat between the end of his GP day job and the start of his evening shift on a seal-diving tour boat and he's texted me to say that he's caught in traffic on the A1. I've spent part of the waiting time in a gift emporium above Seahouses harbour, bypassing the puffin mugs and lettered rock and succumbing to a cuddly seal with weird, staring eyes, freshly branded with a price gun.

Now, I'm at our designated meeting spot at the side of the road sloping down to the harbour, reminding myself what I've learnt over the years about grey seals' biological adaptations to life underwater. The most visible virtue is their streamlined shape, with the bull's penis sheathed inside his body and the cow's nipples recessed, only emerging on land in response to her pup's need for feeding. They also lack external ear flaps and the *arrector pili* muscles that would cause their fur to stand on end, all of which means they can move through water without much drag or resistance.

As I know from having heard many forceful, pre-dive snorts, grey seals exhale before they descend and allow their lungs to collapse, which minimises the problems that can arise from the extreme increase in water pressure. They possess a large blood volume relative to their size, as well as a higher percentage of red blood cells containing more oxygen-carrying haemoglobin per cell than most terrestrial mammals, all of which contributes to their ability to execute lengthy dives without needing to breathe. An additional oxygen-storing protein, myoglobin, is present in their muscles at a high concentration too. Oxygen is further conserved by the fact that blood is pumped just to their vital organs – mainly their brain and heart – when they're underwater, while the latter can slow from its resting, at-the-surface rate of 100–120 beats per minute to less than ten. The risk that they'll suffer from decompression sickness – the bends – is diminished as the dearth of air in their lungs ensures that excessive nitrogen won't accumulate in their blood and bubble into their tissues and joints when they return to the surface.

As far as their subaquatic vision's concerned, large, light-sensitive eyes permit them to see well in murky water and their ability to navigate and locate food is assisted by their extraordinarily responsive whiskers. Tufting from above the eyes as well as much more prolifically from the sides of the muzzle, these *vibrissae* are endowed with around ten times the number of nerve endings as those of many land animals. They also have a uniquely wavy shape, bulging at intervals along their length like broad beans in a pod, enabling seals to detect the slightest displacement of water by fish swimming at a hundred metres' distance. Having observed blind seals looking healthy and well-nourished in the wild and read about studies involving captive seals who, when masked and fitted with noise-cancelling headphones, were still able to detect the subtleties of movements of artificial prey, I'm awed by this whiskery sensitivity.

'Can I borrow you for a minute? I've got to go and pick up a tank.'

Ben's pulled up next to me in the car that he told me to keep a look out for. At first, I stand there in confusion with an image of an armoured vehicle ploughing through my brain: as a non-diver, I foolishly don't grasp the fact that he's referring to a cylinder of compressed air. By the time I've sussed it out, I'm buckled up in the passenger seat and we're heading out of Seahouses with Ben giving me a potted history of his diving-with-seals years from behind his large sunglasses as he drives.

Degree in Oceanography and Biology. A period of time in the Army Reserve including two tours of Bosnia. Trained to become a GP. In 1998, started to explore the waters off the Farnes. The first time a seal swam underneath him he was 'pretty horrified actually'. Became more interested thanks to the encouragement of his wife, a fellow diver, who bought a cheap underwater camera and persuaded him to use it. 'And it just went from there really. When *Springwatch* saw the videos, they said oh my god, that's amazing, we need to lend you a camera. So I ended up filming for *Countryfile, Springwatch, Country Tracks, The One Show*. I became the go-to guy for seals.'

It's a role he clearly relishes and he outlines the reasons why. 'First, there's the scientific side, getting access to things that haven't been filmed before like underwater mating. Plus I've got a stressful job and when you're underwater with intelligent marine mammals who are choosing to interact with you, time just stands still. I've been diving here intensively for twenty years now – I log hundreds and hundreds of hours at the same site every year and I never get bored. I've learnt so much from seals – they've taught me how to dive with them in a way that makes them feel very comfortable. It's to do with real subtleties – breathing, moving, your position in the water.' He lets out a sequence of soft *hoo* sounds from the back of his throat to demonstrate a vocalisation they particularly seem to like. 'I've had seals choosing to lie beside me and fall asleep too – that's a sign of absolute trust and it's the biggest privilege.'

We've pulled up at a storage unit and Ben springs out of the car to pick up the supplies he needs. While I still feel uneasy about his seal-diving activity and specifically its promotion online, I am, at the same time, deeply envious. His descriptions of engaging with seals in their preferred element, of tuning in to the nuances of their undersea behaviour, are no doubt extra-tantalising because

I'm so thoroughly land-bound. My torturous school swimming lessons came to an end when the chlorine in the pool triggered acute migraines – pain and vomiting and scintillating lights. A few years later, I gave it another go, away from term-time jeers and sneers, in the summer holidays, accompanied by Dad who had himself just learnt to swim at adult evening classes. More like a male guillemot coaxing his chick down to the sea from its cliff ledge nest than a grey seal bull who plays no role in parenting, he'd taken the lessons in order to be able to try to nurture my confidence. However, while he was able to offer support and reassure, he couldn't obliterate the chlorine and I again wound up retching and wretched at the poolside. Though I thereafter strove to reconcile myself to my inability to swim, there have been times when I've felt especially regretful. For several years in ocean-fixated Australia, I shrank awkwardly from snorkelling, boogie boarding, parasailing and every other watery jaunt on which I was invited. And though I committed to an outwardly idyllic canoeing expedition with a lover in the wilds of Canada, I spent the whole week privately terrified. Now, living on the coast, I face the sea with a mix of longing, fear, fascination, and envy of those who enter it with insouciance and ease.

When Ben returns to the car, he adds the tank to the other scuba paraphernalia littering the back seat, gives the seal-diving boat operator a call – 'Hello, mate, it's me. I've just nicked a fifteen litre. See you there.' – then starts driving us back to Seahouses, picking up our conversation exactly where we left off. 'It's a real privilege when they play with you too. Something the young males and females do is own you like a toy. One will come in and take ownership of you and if another seal tries to move in they'll chase it off.'

'Have you ever felt threatened?'

'Never ever. They're incredibly gentle animals. Seals have peeled my hood off, taken off my mask, tickled my skin with their whiskers but they've never hurt me – they know the difference between me and my kit. I trust them more than I'd trust any domesticated dog. Honestly, I have no problem putting my hand in seals' mouths when they're in play mode as I know they're not going to bite me. Have you got a dog, Susan? Well, I bet you put your hand in his mouth because you trust him.'

Seahouses is much less frenetic now that it's the early evening and we turn onto the road leading to the harbour in the hope of finding a parking spot close to the departure point for his boat. A woman waves as we trundle down – the mother, he tells me, of one of the boat operators.

'Are you all right there, Ben?' she calls as he zaps the window open and waves back.

'I'll be all right in about an hour when I'm underwater with those seals!'

For the next twenty minutes, we continue our conversation at the open boot of the car while he gathers his gear, answers his phone, greets every passing local, eats some squares of Dairy Milk and flicks a bug off my chin because it looks 'scary'. I assume that his underwater self's a lot less hectic, that entering the sea and spying his first seal cues him to chill and unwind.

'Have you learnt to recognise and ID any of the seals you dive with?' I ask.

'Susan, I can count on one hand the times I've seen the same seal again in all the years I've been diving.'

His response directly contradicts the impression given by some of the news stories that I've read online in which he's represented as encountering the same seals on repeated occasions, leading to the suspicion that they're becoming conditioned to his presence.

'My whole focus is on non-disturbance. These are not tame or habituated seals. And when a seal chooses to come near me, it's a hundred per cent in control of that interaction. Seals in the water and on the land are very different. On land you're a lot more in control and if approached they could become aggressive. Any encounter needs to be on their terms.'

He breaks off again to wave at a car slotting into a nearby space – it contains four of the divers who'll shortly be joining him on the boat. Tonight, there'll be ten in the group; sometimes, there can be as many as sixty.

'Often the people who come to dive with seals may not actually realise when there's one very close by. Seals know exactly where your blind spots are, where your visual field is. Scientists don't like to give that amount of credit to them, to their intelligence. I've had researchers telling me I'm too huggy but I respect the seals and I've been paid back tenfold.' His phone rings again and he glances at the screen. 'It's my sister. I'll be one minute, yeah? Hello, little sis!'

With his kit fully assembled – the wetsuit, flippers and external supply of air all enterprising human attempts to approximate the undersea biological adaptations of the seal – I'm conscious that we're running out of chat time. While I've come to recognise that he himself isn't at all guilty of causing harassment to seals, I still need to ask him about the tabloidisation of the videos he posts online. Don't all the 'Seal Smothers Diver In Kisses' headlines run the risk of generating disturbance by others?

'Well, I'm a lot happier with that than "Seal Attacks Diver" or "Seal Eats All My Fish",' he says. 'I put a clip on Facebook and it got seventy-nine million hits. Seventy-nine million! I was doing interviews to Australia, Japan. . . I just want to use it as a platform – if it comes across that they're really gentle animals rather than vicious, fish-eating monsters, I'm happy.'[2]

17

Industrial Strength

I can't imagine a more surreal seal site. It's like finding Dad's wallet in the bread bin or his glasses case in the fridge. I've travelled about eighty miles south of the Farne Islands and am standing on a bridge over a creek at the side of a main road in the heart of industrial Teesside. And there, on a mudflat just beyond the bridge, in this landscape of belching chimneys and storage tanks, a nuclear power station and petrochemical plants, is a slouch of some eighteen seals.

It's extraordinary that they're using one of the most heavily industrialised areas in the UK as a haul-out. Extraordinary, too, that both greys and commons are resting here in two separate groups. The commons are lying in a looser formation, leaving a generous gap between each individual, while the greys are, as always, tightly bunched. Their urge to lie in close proximity to each other on a warm July day when there's no thermoregulatory advantage to be gained and ample space available in which to spread out, is curious, especially with all the hissing and howling that ensues when a flipper so much as brushes another's flank.

There are apparently two places in the area from which seals can be viewed. I tried to find the other, known as the Seal Sands lookout, which is located nearer the mouth of the estuary, by road earlier, and got lost in a maze of pylons, pipelines, a decommissioned oil rig and fencing fortified by coils of barbed wire. When I at last came to a construction site with signs on its surrounds claiming 'Construction is a Career Like No Other', I cottoned on to the fact that the Seal Sands lookout is inaccessible at the moment. The Environment Agency Flood Alleviation Scheme Phase II is evidently underway, so I had to settle for this viewing point on the bridge over Greatham Creek instead.

It isn't the most comfortable spot from which I've ever watched seals – I'm persistently getting buffeted by traffic as it zooms past – but I at least seem to have arrived at just the right time. The creek is rising and several more grey seals are swimming briskly upstream on the incoming tide, joining those who are already hauled out on the mudflat. They seem skittish and shy, crash-diving and continuing underwater as soon as they spy me. I play a version of Poohsticks with them for a while, starting off on the seaward side of the bridge, then dodging the cars and crossing to the other side to see who's first to surface. They seem to feel safe from my gaze once they've passed under the bridge, following the creek as it winds like a capital S towards the mudflat.

Seals are believed to have frequented Teesmouth for centuries but, by the late 1800s, numbers had drastically declined. Industrialisation of the estuary was, of course, a major contributing factor, as were the related issues of habitat loss, dredging to accommodate larger cargo vessels and persecution by fishermen who blamed seals for the plummeting population of polluted salmon. As a result, by the 1930s, seals had totally disappeared from the area.

According to the most recent Tees Seal Research Programme Monitoring Report, there was, from the early 1970s, a co-ordinated effort by regulators and industry itself to reduce the amount of pollution expelled into the estuary. Gradually, seals began to reappear and there's now a breeding population of commons, with twenty or so pups born at the Seal Sands site at around this time each summer. As a result, Teesmouth is, the report claims, the only

European estuary to have been recolonised by common seals as a result of environmental reform.

Positive though their return to, and year-round presence in, the estuary undoubtedly is, I'd still wager, in the light of what I learnt about Kimberley Bennett's research on the Isle of May, that the seals here are continuing to carry a sizeable toxic load. Since my conversations with her and Kelly Robinson, I've been reading more about the impact of contaminants on both seal and human health. As far as the latter's concerned, scientific findings are increasingly suggesting that exposure to chemical pollutants and their accumulation in the body can have a neurodegenerative impact and contribute to the development of dementia – albeit not the vascular type with which Dad's afflicted but the more prevalent Alzheimer's disease. It's as yet less well known that pinnipeds – based on a study of four seals, five sea lions and one walrus – also appear to manifest features of Alzheimer's in their brains. Most nonhuman mammals develop only one of the two characteristics that impair the brains of human Alzheimer's patients – either the plaques in the spaces between the cells, as is the case with dogs and bears, for example, or the neurofibrillary tangles within the cells, as with cats and leopards. In the brains of the ten deceased pinnipeds that were examined, however, both plaques and tangles could be identified. It's not known if a contaminants burden directly contributed to their disease but it seems feasible to suggest there could be a connection.

Here at Teesmouth, though the small pupping colony of common seals has become established, grey seals choose not to breed but this is due to the fact that the tidal conditions, rather than the pollution levels, are inappropriate. Common seal pups, who've already moulted their white fur and acquired their waterproof pelage while still in the womb, can swim with ease within hours of birth and are therefore able to breed in estuaries where mudflats and sandbanks are exposed for only a few hours a day. As I know from witnessing whitecoats' premature swimming incidents in

Pembrokeshire, however, it's safer for grey seal cows to give birth above the high water mark where their pups can moult their natal fur and grow their adult-type coats.

Because the annual rhythms of the two species vary so much – during greys' on-land breeding and moulting seasons, commons spend more time at sea, while in summer, when greys are mostly in far-off foraging mode, commons are pupping along the coast – there's a greater opportunity for them to thrive in their respective niches. However, since the second phocine distemper epidemic of 2002, common seal populations have continued to decline in parts of the UK and it's possible that a decrease in prey, due to over-exploitation and climate change, could be a factor. Diet studies show that there's substantial overlap in the two seal species' favoured choice of prey – sandeels and gadoids – and it's been proposed that greys may be contributing to commons' decline through competition for these limited resources.

In the next few hours, as the tide continues to flow, the only interspecies competition is over the dwindling mudflat space. The greys become even more tightly and grumpily bunched, while the commons banana themselves into their familiar heads-and-rear-flippers-raised-above-the-water shape. Straggles of seaweed reach the bridge and the row of concrete blocks, former bridge supports that are still in position in the middle of the creek, come close to being submerged.

I assume that some of these seals may occasionally rove further upstream – both up this comparatively small watercourse of Greatham Creek and the adjacent River Tees. Throughout Britain, the surfacing of individual seals in one or other of its landlocked counties is periodically guaranteed to generate a welter of media attention. Almost always reported as being 'fifty miles from the sea' and spotted by a 'startled dog walker', these 'rogue' seals are simul-taneously loved by some locals and loathed by anglers who believe they're stripping the river up which they've chosen to venture of every last fish. In recent years, grey seals have appeared in various rivers including the Trent in Nottinghamshire and the Thames as far west as Teddington while the most notorious upriver swimmer, given the name Keith before experts confirmed her sex, frequented the Severn around Bewdley and Stourport in Worcestershire between 2012 and 2015. Her fans fêted her with two Facebook pages, a Twitter account and a petition arguing that she should be permitted to stay in the Severn instead of being relocated or shot.

The shock of seeing a seal swimming far upstream in a land-locked county would nevertheless pale in comparison with today's experience of watching both commons and greys in this hyper-industrial landscape. Even if I stayed here through umpteen turns of the tide, I don't think I'd ever grow accustomed to the weirdness of viewing a haul-out in the vicinity of a nuclear power station, oil terminals and a gas processing plant.

As the final seal cedes to the risen water and starts to swim and dive, I make my way down from the bridge. From here, I'll be travelling to a more orthodox seal location where, in order to avoid getting stranded, it's even more important to keep in mind tidal lows and highs.

'Everyone says we've got to encourage interaction to make people care about the natural world, but we've created a monster because now people want to interact at all costs.'

I've reached the last stop on the North East of England section of my seal itinerary, and it's also my last stop before I head back to Wales to see Dad and find out if all is as well as our recent phone calls have suggested. From Teesside, I've headed north to Tyneside, then carried on beyond Newcastle for a few miles to the seaside town of Whitley Bay. A ten-minute walk across a tidal causeway, which would have been shorter without all the peering-in-rock-pool pauses, has brought me to tiny St Mary's Island, topped with a lighthouse like a candle on a cupcake. And here, on the North Sea-facing side of the lighthouse, I've rendez-voused with Sal Bennett, one of the founders of the islet's Seal Watch Group.

We're at the lighthouse perimeter wall, beyond which are rostrums of rock, skeins of kelp, and the sea, currently about sixty metres distant. I count twenty seals, all greys, mostly adult cows, lolling at the point where water meets rock. It's a breezy day – Sal has her St Mary's Seal Watch fleece zipped to the chin and keeps having to swat blonde strands of hair from around her mouth as she speaks.

'Five or so years ago, before the group started, there'd be maybe six seals at most and that would only last until someone shouted "Seal!" and they'd all leave. We could see they were really struggling and wanted to try and reduce the disturbance. To begin with, we just tried to create a bit of a buffer – asking people if they'd mind not going right up to the seals and, over time, we increased the distance. Now we use this lighthouse wall and ask people to keep behind it.'

'Have seal numbers grown a lot since then?'

'Yes – it's a spring and summer haul-out, not a breeding site, and peak numbers last year were in the mid-fifties. We've got enough information to know that it's not just new seals dropping by too – some of the adult females here now are the same ones who were coming five years ago.'

Another fleeced St Mary's Seal Watch woman has arrived at the lighthouse wall, followed by a man with a telescope. Tracey, a mature university student who used to breed alpacas in Scotland, and Rod, who describes himself as a 'very local' resident, are volunteers with the group. For the next two hours, they'll be doing visitor engagement, offering information about the seals and encouraging people to keep their distance and do their viewing through the telescope.

'About a year after we started, we had our first training day – that's when we got our first volunteers,' Sal explains. 'We've got about sixty on the books now. Initially, we only came here at weekends but this year we're covering most weekdays too. Like me, the majority of our volunteers – and this is why I love it – aren't from ecological or wildlife conservation backgrounds but people from the community who just got really annoyed when they saw what used to happen here.'

Visitor numbers are low so far today so there's no opportunity for me to witness the persuasive powers of Tracey and Rod. 'Are people generally quite receptive when you suggest they shouldn't disturb the seals?' I ask.

Sal nods. 'Many are. We use peer pressure – once you have a certain amount of people doing something in a particular way, it makes it more obvious that there are a few who are doing it differently. Our biggest problem is with people who've been coming here for years and getting them to change from how they've always used the site – you hear things like "I've been fishing here for thirty years and my father and grandfather used to come here too." But we also get visitors who really value the site and are very protective of it. That's what we want – people taking responsibility. We need people to take on ownership and then the group can back off a bit.'

At Sal's suggestion, we back off into the lighthouse, both to shelter from the breeze and to seek an alternative view of the seals. Entry is via the former keepers' cottages that have been converted into a gift shop and visitor centre. As we wander past a What Is Biodiversity? info board, and a display of old lighthouse lights, she asks me which other seal sites I've visited. Teesside inspires a grimace ('That's the sort of place where you really don't want to take a deep breath in') while she refuses to go to the Farne Islands any more ('I find it too disturbing. They're pimped, they're cash cows for the National Trust and with people's insatiable need to interact with wildlife, the boats get much too close'). She's full of enthusiasm, though, for both Ythan Seal Watch ('Lee and his wife came down here to talk to us before they started their group') and Cornwall Seal Group Research Trust ('Sue Sayer's been our mentor from the start').

To get to the top of the lighthouse, we have to corkscrew our way up 137 steep stairs. Keeping Sal's black ankle boots just above my head the whole time we climb, I likewise express my admiration for Sue's monitoring work on the cliffs near St Ives.

'Every site's different, though, of course,' says Sal. 'I've had to apply what I learnt from Sue to make it work at this unique site – an island that's about two hundred metres across, pedestrian access, fairly urbanised, a holiday resort within easy reach of Newcastle. . . It's not enough for it to be managed in a traditional way.'

We pause at the bottom of the final set of stairs, a ladder, so that a couple of visitors can clamber down. 'So who actually manages the site?' I ask.

'The local authority – North Tyneside Council. It's part of a Local Nature Reserve that was designated in 1992 but there's still only a draft management plan – a *draft*! When we were first setting up the group, the council didn't even know it *was* a nature reserve. It also includes a wetland and grassland area at the north end of Whitley Bay beach that used to be very vibrant. Now, though, you're more likely to see dog poo there than a flower. As usual, it's the wildlife that's being compromised.'

When we emerge from the ladder steps onto the enclosed viewing platform at the top of the lighthouse, she points, with a purple-nail-varnished finger, towards the low cliffs of the mainland to show me the extent of the reserve, after which I swing round to check on the seals. Our gannet's-eye view makes it easier to get an accurate sense of numbers – a sedate group of eight is cushioned on kelp while a more restless gathering of sixteen is suffering the lumps of the rocky substrate. Between the two groups, several cormorants stand sentinel.

'So has it been challenging, at times, working with the council?' I ask.

'Yes, but we've been stubborn enough to persist. And we've also done it in a way that the local authority didn't anticipate. We're quite gentle – we didn't come in kicking and screaming and demanding. And we worked our way in, most importantly, with the support of the community – without that, we wouldn't have any influence at all.'

Having watched the haul-out for an uneventful while, we start spiralling down to ground level again with Sal telling me about a recent development bid to revamp the lighthouse. 'The council wanted an extension with a rooftop terrace and we knew it would be the end if that happened – can you imagine the disturbance to the seals? Lots of people, including other conservation groups, said we shouldn't challenge it and work with the council instead but I thought no, why should we? And it got rejected by the planning committee – there were something like six hundred objections!'

'So the council had to submit new plans?'

'Yes, and considering we're just a group of people from the community, that was massive. We're an absolute zero-budget organisation too. All our equipment – telescope, binoculars – has been donated. And I never really had a Big Idea, just that it wasn't right that the wildlife always, always, had to accommodate people. And everything grew organically from that.'

We've reached a room on the ground floor of the lighthouse complex furnished with a rocking chair and a stack of green plastic seats, and with a picture of a shipwreck and a marine life mosaic on the wall. Sal settles on one of the green seats and indicates I should take the rocker – its motion offers a soothing counterpoint to her account of some of the worst disturbance incidents that the seals here have suffered.

'We got footage of some youngsters throwing rocks at them. We'd had reports of it happening before but this was the first time we got it on camera and had enough evidence to report it to the police. I don't think they meant to be violent, I think they just did it to get a reaction and make the seals move because if you look at it from a purely comedic point of view, if you're not engaged at all with what's actually happening, seals' movements on land are quite funny. I suppose they film it and put it on WhatsApp or something.' She goes on to convey that the large tidal range here means that the seals get stranded on the rocky terrain when the sea falls and while disturbance doesn't cause a stampede on the scale of that at the Ythan Estuary, they panic-scramble and tombstone – throw themselves into the water or onto lower-level rocks from quite a height – with the risk of critical injury.

From her cotton tote bag, patterned with Russian matryoshka dolls, she withdraws a St Mary's Seal Watch leaflet, in which this risk is spelt out and a Voluntary Code of Conduct on how to behave around seals is included. The tone is pleasant and reasonable – 'As a visitor, please be willing to flexibly change your plans to give seals the space and time they need' – and the rationale behind this request is lucidly expressed too, through information about seals' feeding and digesting routines and the impact of disturbance on their energy levels and health. I'm particularly taken by the advice on how many metres visitors should leave between themselves and the seals. Instead of issuing, as most other sites do, an abstract 'at least fifty metres', helpful context is offered – 'As a reference, the lighthouse tower is 38 m high.'

There's further information about best seal-watching practice on the walls of the wildlife hide to which Sal and I wander, at a short distance from the lighthouse. This time, the tone seems deftly designed to make visitors feel good about their behaviour and therefore more willing to sustain it. 'Thank you for viewing the wildlife responsibly from inside this hide. Your respect and consideration will help the wildlife remain on the rocks undisturbed, ensuring a better future for it and an opportunity for others after you to see these wonderful animals in their natural environment.'

Having hoicked myself up onto an uncommonly high bench fronting the hide's row of windows, I notice there's also a small display about birdlife on one of the walls – the island is frequented by a range of species including sanderlings, turnstones, lapwings and purple sandpipers.

'Yes,' Sal confirms. 'This site isn't just about seals. One weekend last year when we did a bird count, we had over a thousand golden plover pouring out of the sky as the tide came in.'

For now, there are still only cormorants in view, their lanky perching pose and the flock of offshore wind turbines the only uprights between here and the horizon.

'You know, people see cormorants sometimes and ask me if they're penguins,' Sal says suddenly.

There's something about the confines of a wildlife hide, perhaps the intimacy generated by low voices and low light, which often seems to inspire unexpected revelations. I've already got a feel for Sal's steeliness and quiet persistence in setting up and progressing St Mary's Seal Watch, but now her sense of the absurd comes to the fore as she delivers a monologue about some of the more outlandish situations and visitor comments she's had to deal with.

'We have a very, very strange relationship with seals – people think they have to drag them back into the water. . . There's a lot of ignorance, considering they're an indigenous species and we've lived alongside them for so long – it's like an alien's landed. . . We get some really weird ones, like the woman who tried to insist she needed to walk near the seals because otherwise it would mess with her feng shui. . . I once heard a man say, "Is this where they hatch?" . . . The other day we had someone asking if we take them back to the Sea Life Centre at night and what time we feed them. . . Then there was the year that *Springwatch* made a short clip – the following week we had a Page Three model here. It was Father's Day, very busy – she had a tiny bikini top on and two guys were photographing her on the rocks in very explicit positions. . .'

My own position on the high, narrow bench, legs too short to reach the bar of the footrest, is explicitly uncomfortable. By contrast, the fidgety group of seals on their breeze-buffeted bed of rock has subsided into stillness. Hopefully, they'll remain there undisturbed until the tide comes in and floats them off again. I won't be able to witness this, though, as I'll need to cross the causeway before the sea does, access the mainland and start on the long journey back to Wales.

Before I go, I have a final question for Sal – the one that asks her to consider what she finds most fascinating about grey seals. Given that she's engaged with the challenges of this unique site with such imagination and originality, I'm fully expecting to get a different kind of response from those I've received from other seal group co-ordinators and rangers, who've raved about whiskers, eyebrows and pups' swimming capers. But her reply still takes me by surprise.

'To be honest, I wouldn't say I'm fascinated by seals. It's the way humans behave around the seals – the conflict – that makes them more interesting to me.' She gives a slight shrug of her shoulders and smiles. 'As an animal, I find them no more fascinating than the hummingbird hawk-moth that we had here last year.'[1]

18

Elders and Betters

The coast path has disappeared in places, overwhelmed by bracken and brambles, and foxgloves on the wane. With just the top third of their blossom remaining, they look like giant jingle-bell rattles waiting for a shake. Lower, alongside Hooper, the track is a ruffle of common toadflax, and blue bobbles of sheep's-bit brush my shins and his nose. The scent of wild honeysuckle is weaving through the warmth of the early evening while Hooper's ears veer first towards the chirr of grasshoppers, then the whirr of a golden-ringed dragonfly's flight. After several months of travel up the northwest and down the northeast coasts, plus the in-between weeks spent with Dad after his mini-stroke, I'm back at my home bay of Aber Felin where summer means abundance for all species but the seal.

Speckled wood butterflies scatter as we thrash through the bracken to reach the viewing point over the first beach. So populous with pups and cows in the breeding season, it's never used as a haul-out at this time of year. For some reason, the few seals who frequent the bay from April to July seem to prefer the rocks off beach number two, so we wander on to see if they happen to be occupied today. Towards the horizon, the sun is scintillating on the sea's surface like a migraine aura while closer to the cliffs, the water is uncharacteristically clear and turquoise like it's been piped in directly from the Med.

Though there aren't any seals on the rocks, I spot three young cows cavorting underwater, zigzagging, somersaulting, sometimes flipping over to swim on their backs. If the sea were its usual grey, rumpled self, I'd never have known they were here. As I watch their subaquatic play, I wonder if they're regular or occasional visitors or if they're discovering the bay for the first time. I've made little headway with the ID catalogue that I planned to set up after meeting Sue Sayer so my photo library won't be able to offer any immediate insights. I wish it were possible to know where the bulk of Aber Felin's breeders and moulters disappear to over the summer months too, and which feeding grounds they favour when endeavouring to build up resources for the pupping season. Is my supposition that some travel north to the West Hoyle sandbank correct? How widely do they move around the Irish Sea and do any make more extensive journeys to Cornwall or Brittany?

At last, the most petite of the cows seems to tire of both play and bay and decides to swim away. I follow her progress through my binoculars, something else I wouldn't be able to do in more typical weather and wave conditions, noting how she gains momentum by moving her back flippers and the lower part of her body from side to side, such a contrast to the wing-like, foreflipper-flapping propulsion of sea lions. Grey seals have the capacity to swim at thirty-five kilometres an hour but she's moving in a more leisurely fashion, evidently adopting the more usual cruising speed of between four and ten. As she disappears around the eastern headland, the largest cow stops capering too and drags herself onto one of the rocks. The remaining seal snorts and gurgles, the sound of a child blowing bubbles in a drink with a straw, then tries to resume the game by swimming round the rock and launching herself onto it from behind. The other cow pivots, howls and flaps her front flipper, though, unequivocally signalling the close of play.

When we resume our walk, Hooper lingers at a hollow that's functioning as a badger latrine, diligently sniffing both the dark splats of shit that have been deposited there and the white guard hairs snagged on an adjacent twist of barbed wire. While he's preoccupied, I enjoy the kaleidoscope of butterflies – red admiral, meadow brown, small copper, common blue – and try to appreciate the flowering grasses too. Though towering Timothy and the purplish heads of Yorkshire fog are pleasant enough, I find it impossible to delight in sweet vernal grass, however enticing its name. A few years ago, Hooper freakishly inhaled one of its spiky seeds: it became embedded in his throat, causing extreme swelling that severely restricted his breathing. Without emergency surgery, it would have cost him his life.

The third, most sheltered beach is empty of seals today too so I turn my attention to finding an appropriate picnic spot. One option is a mini-headland from which there's a view both out to sea and back towards a section of cliff that hosts a small colony of nesting fulmars. I've enjoyed watching the progression of their breeding season for several years now – the fact that they can live beyond the age of forty and are faithful both to their nest site and each other means I can feel pretty confident that each spring and summer I'm viewing the same twenty pairs of birds. I've seen their bill-clacking, cackling courtships; a protracted mating – he vigorously twisting, she rather flattened and passive; the plucking of grass and strands of thrift and kidney vetch from the cliff to furnish their nest-scrapes; and the shared, patient, fifty-day incubation of their egg. Throughout all these observations, I've fortunately managed to avoid being perceived as a threat as, in order to defend themselves, fulmars projectile-vomit rank, acidic stomach oil. A friend who was sprayed when leading a bird tour in Iceland still speaks of the horror of having to burn all the clothes he was wearing as no amount of washing would remove the foul stench.

Today, I'm hoping I might glimpse some chicks, resembling, at this stage in their development, overgrown dandelion clocks. However, as Hooper and I advance onto the headland, I notice a movement that causes me to emergency-stop and pull back sharply on his lead. An adder's sliding sidelong from its basking spot, the rock on which I planned to squat, into a clump of bracken.

Hoping that the next headland, a low, rocky promontory, like a gnarled finger jabbing at the Irish Sea, won't be similarly off-limits to picnics, we walk the short distance to reach it. I missed its succession of spring flowers this year – scurvy grass yielding to bluebells, spring squill and English stonecrop. Now, as we drop down off the main path, there's a cushion of wild thyme pinking into bloom.

At its yellow-lichened tip, the headland's barely a metre wide so we opt for a broader, grassier spot about halfway along. After giving Hooper water and kibble, I open my Tupperware supper of salad and quiche. We're much closer to sea level here and the little bay over which we're looking is dotted with lobster pot markers, some round and orange, others rectangular and black. I think I glimpse one that's more eccentrically shaped but then I realise it's a barrel jellyfish, frilled arms flaring out like drowned Ophelia's hair in Millais' pre-Raphaelite painting.

Having polished off his food, Hooper turns his attention towards mine, then flops onto the grass with a resigned exhalation when he recognises that its main component is spinach. Once I finish eating, I rinse out the tub with water from my bottle and leave it to dry in the sun so it can be filled with elderflowers on the way home. It's late in the season but I passed a tree on a farm track earlier from which I've gathered blossoms in previous years and there looked to be just enough healthy flowers left with which to make a small amount of cordial.

First, however, Hooper and I both close our eyes and snooze for a while, waking to find the sun lower in the sky, turning the sea to honey. A seal is bottling close to our headland too, not the petite young cow who swam away from the game in front of the second beach earlier but an older one with heftier shoulders and a scar above her right eye. She stays in her vertical position long enough for me to take some photos, capturing the pattern of punctuation marks – three inverted commas and a full stop – mottling the left side of her neck, for inclusion in my ID catalogue should I ever manage to get it properly established.

I suspect I'd be able to recall this seal without the aid of photos, though, her whiskers and eyebrows gold as a harvest, her wet pelt sheened with light.

It's like I've bottled that evening sunshine.

Standing on tiptoe, I stretched up to pick any sweet-scented elderflowers that were not yet browning and gently shook out the bugs. Carried the blossom home in my Tupperware tub and a cotton bag. Boiled sugar and water to make a syrup. Added the flowers, citric acid and lemon zest and left it to steep overnight. Strained the mixture through muslin. Sterilised two bottles and decanted the liquid sun.

Today, I'm delivering one of the bottles of cordial to Dad. I've already seen him since travelling back to Wales from St Mary's Island and was relieved that he seemed as well as when I left for Orkney several weeks previously. He had only a moderate moan about the fact that Bethan's care visit interrupts his plans for the day and apart from there being a smoke alarm battery that needed to be replaced, prompting a repeated 'What's that blasted beeping?', there weren't any household issues to deal with either. I'm hoping that today's visit will be equally serene.

I've brought the ingredients of a ploughman's lunch along as well as the cordial and carry it out all into the garden. Dad and I sit with a view of my rusty childhood swing that now doubles as a bird feeding station, and Hooper lies at Dad's feet.

'You didn't have to bring your own food with you. I've got plenty here, you know.'

A wood pigeon flaps down from the top bar of the swing to the grass beneath it and starts pecking up the seed that's been scattered by the young blue tits on one of the feeders.

'What's that?'

'Wood pigeon,' I say.

'Not the bird. That drink in the glass.'

I explain about the cordial making. He smiles as I describe the tree's coastal location and how the elderflower clusters look like ivory lace parasols, then assumes a more troubled expression as soon as I suggest he samples it. 'But I always have tea at this time.'

'We can have tea as well. Go on – try it.'

For as long as I can remember, he's cleaved to a routine but dementia seems to be reinforcing and magnifying that trait. It comes to the fore again post-lunch when he decides he wants to move back inside and drink his mug of tea in his usual armchair. I stay in the garden a little longer, playing fetch with Hooper who needs a bit more exercise after the journey here, then head inside to do the washing-up before joining Dad in the lounge.

He's dozed off to sleep, chin on the shelf of his chest and mug on the nest of tables next to him.

I sit on the sofa, find a half-completed crossword from yester-day's paper, fill in 8 across, 15 across and 4 down.

Let him snooze on or wake him? He'll be thoroughly miffed if he sleeps my visit away – even more so if he's let his tea go cold before drinking it and finds that I've answered all the clues.

'Hey, Dad.' No response. 'Dad – shall we see if we can finish this crossword?'

He's in a deeper doze than I realised. I rub his arm, but only gently as he's been prone to purple bruising since starting the blood-thinning medication. 'Dad! Wake up!' Still no response. 'Wake up and have some tea.'

I try again with the arm rub, shift to tweaking his cheek but neither my voice nor touch gets any reaction at all.

The cordial. What have I done to him?

Can't be the cordial. I've drunk it too.

'Can you hear me, Dad?' I'm shouting now, right next to his left ear, while simultaneously pushing the collar of his polo shirt aside and feeling for the pulse in his neck.

Slowish pulse. Still breathing. His chest's lifting his chin and I can feel his breath on my hand.

Hooper, sensing that something's awry, plants his front paws on

Dad's lap and levers himself up to lick his face. Canine connection of this kind would usually inspire exuberant affection from Dad in return.

The licking gets more insistent.

No sleep can be this deep.

For the third time in five months I'm on the phone to the emergency services. For the third time my narrative's mangled by panic.

At least this time I'm not doing it from a distance. I can keep a close watch. Try to describe any changes

'I've just noticed his mouth – the right side – no, left – he's just started dribbling—'

'Does your father suffer from seizures at all?'

'No, he's never—'

'No history of epilepsy?'

'Just a mini-stroke a few months – about six weeks – ago.'

'Now listen my lovely, the ambulance is on its way but I need you to get him onto the floor. Can you manage to do that for me?'

'I'm not – what do I—'

'You'll have to pull him down out of the chair.'

'I'll hurt him—'

'He needs to be lying on his side. We've got to keep his airway open.'

Tug one leg, then the other, then back to the first leg again. Heavy but not as heavy as I expected. I wince as his bottom bumps down from chair onto carpet, and battle to prevent his head from getting similarly jolted as I manoeuvre him first to horizontal, then onto his side in something resembling the recovery position.

'Marvellous, well done. Now, the paramedics will be with you shortly. Make sure they can access the property and please shut any animals in a separate room.'

Hooper's lying as close to Dad as it's possible to get, periodically nosing his hand. Dragging him away into the kitchen when the ambulance comes will be even more of a challenge than getting Dad onto the carpet.

I hold the hand that Hooper's been nudging, try to warm the chilled fingertips. I take off his glasses so he won't squash them and gather up the coins that have spilled from the pocket of his jeans. I tell him a second woodpigeon turned up while I was in the garden, as well as a tiny coal tit zipping in and out of the conifers.

Suddenly, a mumble. A series of mumbles. Slurred sounds, nothing approaching words.

'Dad?'

An eye twitch.

'Dad! Wake up!'

From twitch to flicker to half-open.

'Can you hear me?'

He frowns and seals his eyes up again.

'Come on, Dad!' I shake his shoulder, and he lets out more of a grumble than a mumble, trying to shrug me away. I give it another shake, which coaxes his eyes into a squint as he tries, with one side of his face against the floor and no glasses, to bring me into focus. Confusion, then realisation and a whisper of a smile. 'Hello, Ma.'

As soon as he was diagnosed with vascular dementia, I knew I was dreading the non-recognition phase most of all and now I've got a sudden, unexpected preview.

'Dad – can you tell me who you are, what your name is?'

Rapid blinks and a slight shake of the head. 'Fred.'

While I sheer between relief that he's conscious and alarm that he's so addled, Hooper's reaction is unequivocal – up on his feet, wagging his tail, tongue out, joy-panting. Having registered his presence, Dad's frown morphs into a tentative grin and, after a few more minutes of blinking and thinking, he manages a delighted 'Hooper!'

Following this reconnection, he becomes gradually more lucid, questioning his presence on the floor and returning, bit by bit, to himself again. By the time the paramedics, who introduce them-selves as Bev and Mike, arrive, he's complaining about my having insisted he stay resting on the carpet and the fact that his left arm's gone to sleep. They establish it's safe to move him, help him back into his chair and run through a series of tests from the cardiac to the neurological.

As Bev shows me photos of her dogs on her phone, Mike assesses Dad's balance and ability to walk in a straight line, Hooper padding close behind.

'Nice garden,' Mike says, glancing out of the window as they return to the lounge from the hall.

'Grass needs a cut,' Dad replies.

When he's seated again, Bev writes up some notes. 'Now then, Geoff,' she says, 'all your vital signs are fine at the moment and the episode you experienced has clearly resolved. But given your age and medical history, I think we need to get you to hospital so you can be checked out further.'

Dismay glazes Dad's face. 'What do I want to go to hospital for? Never been there in my life.'

This time, Dad's stay in hospital is brief and inconclusive. The tests that were carried out following his TIA are repeated and though doctors are reluctant to definitively label this latest episode as another mini-stroke, they agree that an inadequate supply of blood to the brain is likely to have contributed. He's added to the waiting list for an ambulatory ECG which will require him to wear a portable heart monitor for several days while he goes about his regular activities like the cow seals in the stress response research on the Isle of May. His medication is also slightly modified, with a different blood-thinner being prescribed. In less than a day, he's home again and in less than a week, he has no memory of what happened at all.

By contrast, my memory of the incident has inflated and intensified. At the time, I was so focused on practicalities – phoning for the ambulance, trying to call Dad back to consciousness, liaising with the paramedics – that I was barely aware of the ravel and writhe of my own feelings. Now, though, if I somehow manage to snitch a few hours' sleep, I wake drenched in remembering. The fear that no medical professional, treatment or procedure would be able to rouse him. The conviction that I missed something, could have pre-empted it, should have known. Should have tuned into some tiny change in his behaviour, just as seals sense subtle shifts with their *vibrissae*. I even chide myself for being so enraptured at Aber Felin last week, for relaxing into the fecundity of summer when another crisis with Dad was bound to be looming. Other crises could be brewing now too – when I'm with Dad, I stay on hyper-alert and whenever I'm not with him, I flinch each time my phone rings, needing to draw on every yogic breathing technique I've ever learnt to slow the hurtle of my heart.

In the past when my anxiety's reached this peak, I've grasped at a myriad other remedies in addition to yoga. More energetic exercise. Epic coastal walks. Cognitive behavioural and psychodynamic therapy. Meditation. Homeopathy. Emotional Freedom Technique. And, most unconventional of all as far as western medicine is concerned, shamanic healing.

My introduction to this was unsought, inadvertent. In the year after I returned to the UK from Australia and beyond, and got drawn into a vortex of creative writing teaching for a ridiculous number of organisations and institutions, one of my most inspired and imaginative students was a woman named Sarah Howcroft. After acing every module she elected to take, she told me she'd like to continue her writing journey outside the university setting and disclosed she'd been working as a shaman, having trained and developed a practice grounded in several indigenous traditions, for over two decades. If I was willing to continue as her writing tutor and mentor, she'd be happy to pay me in shamanic healing.

While I approached the initial sessions with a mix of scepticism and curiosity, the former was, over time, supplanted by gratitude and pleasure. As well as receiving Sarah's healing, I learnt to let the drumbeat usher me into an altered state of consciousness and navigate me through journeys to the Lower and Upper Worlds, where I acquired a tribe of power animals including, not surprisingly, Seal. Eventually, I joined Sarah's Medicine Women's Circle and we even ended up running writing and shamanism workshops together. Though I came to understand that anxiety, at some level, would always be part of my psychological and chemical make-up, Sarah's teachings helped me feel more in balance and turn the worry dial down.

In more recent times, however, I've lost impetus. Sarah shockingly died a few years ago within months of receiving a cancer diagnosis, and establishing a connection with another shamanic healer would have felt both distressing and disloyal. Yet in these wobbly weeks in the wake of the latest Dad drama, I start to reflect on whether it might be time to reach out again.

When I was immersed in selkie research a couple of months back, I came across a shamanic practitioner named Olwen, elements of whose approach draw on Celtic traditions and who's especially interested in working with seal energy and selkie myths. We chatted on the phone a few times: I warmed to her manner and, aside from her selkie insights, she urged me to get in touch if I ever felt the need for fresh shamanic input in my life.

If I use my dawn fretting hours for travel, I work out that I'll be able to get to and from the village where she lives and practises in a day. In between this physical journeying, there will, I assume, be a guided 'soul flight' to the Lower World, which will enable reconnection with my power animals, Seal included, and remind me of the benign jolt to the psyche that accessing 'non-ordinary reality' can provide.

I feel as if I'm teetering like the stalks of angelica lining Olwen's drive as I make my way to the garage that she's converted into a shamanic healing and workshop space. It's another sultry summer day and, having slipped off my sandals at the entrance, I'm glad of the cool oak floor beneath my feet. It doesn't help to ground me, however – my attention and gaze keep straying to the window and its distant view of the sea; to the display of beaded shamanic drums; to the altar area with its sand, water, candle and kestrel feathers symbolising the four elements; to the unremarkable beanbags.

Olwen invited me to bring something along to add to the altar, a representation of one of my power animals perhaps. I only remembered this instruction in the flustered moments before I left the house early this morning and grabbed the first thing I saw, the cuddly seal with the vacant plastic eyes from the gift emporium in Seahouses. It had somehow found its way onto Hooper's toy pile and I feel embarrassed about elevating it to the altar – its weird blue stare and grubby white fur give the impression that I'm failing to take this session seriously.

After the ritual smudging – burning a sage stick with the aim of clearing the space of negative energy – Olwen asks me to articulate what's brought me here and what I'm searching for in terms of guidance and healing. I want to stop feeling like I'm entangled in an abandoned trawl net, I say. I want to rediscover the ability to sleep for more than three fitful hours per night. I want to learn to accept Dad's health conditions and be more at ease with expecting the unexpected. And I want to be able to complete my seal odyssey – I originally planned to return to the east coast of England when the autumn pupping season gets going but anxiety's telling me that's impossible now.

Olwen's recommended remedy initially feels impossible too. We'll do some Lower World journeying, yes, but I might gain greater benefit from a more radical intervention. To maximise the healing that my power animal can offer, and in view of the fact that I have previous experience of shamanic teachings, I might consider letting her guide me through a deeper process, often facilitated via a trance dance, which will afford me the opportunity of merging with Seal.

'But as you know, shapeshifting of this kind mustn't be undertaken lightly. We'll need to set clear boundaries.' She briefly switches from the smiling vibrancy that reminds me so much of Sarah to expressing words of caution. Once the spirit of Seal has been summoned to take up temporary residence in my body and I've explored that experience and received guidance, the letting go, the requesting Seal to leave must be carefully and thoroughly enacted so that he and I are separate again.

Rather than summoning Seal, I feel like I've summoned all my old scepticism. After a distracted, unsatisfactory Lower World journey and a break for herbal tea, I'm standing with my eyes closed, trying to deepen my breathing. Olwen, on the other side of the altar, has started to beat her drum, creating the rhythm that's supposed to persuade my brain to enter an alternative wave state. However, my brain's currently busying itself with working out ways of avoiding the camera she's set up to capture this whole shapeshifting spectacle. Viewing the visuals afterwards will apparently be helpful to me but it's making me feel self-conscious in the extreme.

'I call upon Seal to join our circle and enter the body of Susan,' she announces loudly and emphatically.

The urge to laugh surges to my surface, threatening to breach with the force of a humpback whale. Shamans in indigenous communities have been performing shapeshifting rituals for millennia but this is just play-acting. Shallow make-believe.

'And I invite Susan to call upon her power animal too.'

Quick flick back through the selkie tales and novels I've read. Should I find a phantom sealskin? Pretend to wriggle in? 'I call upon Seal. . .'

'Wow!' says Olwen. 'That was fast!'

I'm not exactly sure what she means but it seems I've swapped standing for being belly-down on the floor.

The drumbeat tweaks my feet, fans them out like flippers.

It propels me along on my forearms through heat so thick it feels more like water than air.

It welds my right leg to my left, melds calves, knees, thighs. Quells the obligation to be upright.

Beanbag is rock.

Floorboards are beach.

I let go of words, haul out in the language of scratch and howl and uncompromised sleep.

In the time it takes the tide to turn – or in an hour or a minute or a week – a change in the rhythm of the drum calls me back to the converted garage.

There's the candle.

There's the altar.

There's my half-drunk mug of peppermint tea.

And there's Olwen by the door, laying her drum and beater on the floor. Her grey hair's waterfalling over her shoulders and she runs both hands through it, lifting it to cool the back of her neck.

I make circles with my own neck – gone is the bulky sense of blubber – and give an experimental flex of my feet, testing for the stretch of webbing between my toes.

Olwen nods her approval. 'Good. Remember what I said. We need to make sure you're entirely you again.'

As part of our debrief, she downloads the contents of the camera but what I see on the screen bears little relation to how I was feeling inside. Was I outwardly still so resoundingly human? Was I really still wearing my glasses and watch?

My face I barely recognise. It looks so different from the face that mopes back at me from the bathroom mirror. No brow ridge and furrow. No jaw clench. No moonless midnight under the eyes.

Whether or not the drum nudged me to invent the presence of Seal, whether or not my sensation of shifting shape was a sham, I was briefly yet completely fret-free.

19

Sonic Boom

'Woah! What d'you reckon? Eurofighter Typhoons?'

A trio of fighter jets has erupted from the low clouds to the north and they're screeching towards the seal colony, streaking over sandflats and salt marsh and zooming in the direction of the dunes. While my instinct is to dive for cover, the other visitors lining the fence that separates us from the seals collectively swing their phones, tablets and jumbo lenses from suckling pups to sky.

Astonishingly, the immensely pregnant cow in front of whom I've been stationed all morning here at Donna Nook Nature Reserve in Lincolnshire, in the hope of observing my first birth, barely twitches a whisker. The bull seals, mud-wrestling in the marsh in the middle distance, also seem oblivious to the noise and carry on fighting for mating rights. Closer to the fence, however, the non-habituated newborns cower and flinch, while some of the more mobile moulting pups propel themselves in frantic circles. Their mothers track their movements and try to give nose-to-nose reassurance.

One of the volunteer seal wardens, noticing my disturbed expression, shrugs his shoulders. 'Got to practise somewhere.'

In the course of the next half hour, the jets roar over the colony twice more and though I'm newly alarmed each time, the pregnant cow still isn't shocked into giving birth. Throughout the morning, she's displayed all the signs that she could be going into labour – fidgeting, digging with her front flippers, spreading those at the rear – but keeps being overhauled by sleep instead.

It's a stingingly cold day in late November and I've finally resumed my grey seal journey by heading all the way back over to the east side of the country. Pupping in Pembrokeshire peaked in late September and I spotted the last weaners on the beaches at Aber Felin at the end of October. However, since the start of the breeding season varies according to location, progressing in a clockwise direction around the coast of Britain, the east of England colonies are only now starting to build towards maximum numbers.

As my edgy reaction to the fighter planes shows, my creek of anxiety continues to flow, albeit with much less momentum than in the weeks following Dad's loss of consciousness. His health has been stable since then, the shamanic session may have had some lingering benefit and I gradually came to believe it would be possible to recommence my travels. He's cognitively sharp at the moment too: when I phoned last night to check in with him and let him know that I'd arrived, there was none of the usual confusion about my location. Since he was born just a short distance up the coast from Donna Nook in Grimsby and has retained strong long-term memories of his Lincolnshire childhood – the Anderson shelter in the back garden, burying his ice cream cone in the Cleethorpes sand to deter the wasps, falling off Pink the pony – he can't wait to hear what I'm going to tell him about his native county eight decades on.

Seals don't feature in his Lincolnshire memories at all as they only started to appear at Donna Nook in small numbers in the 1970s. The colony's rapid expansion in subsequent decades seems extraordinary given that the pups share their nursery with RAF Donna Nook, the source of the low-flying fighter jets. While Lincolnshire Wildlife Trust manages the reserve, it's owned by the Ministry of Defence and though certain bombing drills on the sands beyond the marsh are paused during the breeding season, other, modified forms of target practice are ongoing. There's a remarkable variety in global grey seal pupping habitats – they've adapted to the darkness of caves, the jags of rocky coves, dunes and huge sandy beaches. A number of cows on Hebridean North Rona slog their way up a steep, eighty-metre-high hill in order to claim a birthing spot and populations in the Baltic and Canada's Gulf of St Lawrence even breed on ice. Yet this proximity to the military is unprecedented.

'A special place where wildlife lives in harmony with the military planes' is the positive spin given by the Welcome to Donna Nook information board in the reserve's car park. I arrived here early this morning at the same time as the Tasty Treats snack van – 'Serving Donna Nook Since 1997' – which, together with the yellow All Seal Traffic road signs directing cars to the coast from the adjacent village of North Somercotes, suggests that this will turn out to be the busiest seal site I've visited.

It consequently feels like the most managed site I've visited too. While the colony stretches for several miles, public access to this innermost section is legitimately limited to a footpath behind a double fence along which a plethora of instructive signs is fixed. A warning to keep dogs away as grey seals are carriers of the phocine distemper virus which may be transmitted to canines. Placards about the birth process, mother-pup bonding, different seal sounds. A kids' multiple choice quiz ('Does an adult seal have 34 whiskers, teeth or swimming certificates?'). And a glut of look-but-don't-touch advice. ('You may be tempted to stroke a pup lying against the fence. Please don't. Your smell on the pup may cause the mother to abandon it.' 'You may be tempted to encourage an inactive seal to "do something!" Please don't. The seals need to conserve energy and are not here to provide entertainment for human visitors.')

Micromanaged and militarised though the site may be, it's the most extraordinary I've yet been to in terms of the observation opportunities it offers. Instead of watching pup activity through binoculars from high on a cliff top above a beach, I'm the width of a double fence away. I'm close enough to see milk dribbling from cows' teats, snot sneezed from newborn noses. I can hear pups' fizzling farts and the massively pregnant cow's raspy sleep breathing. There's the wet slap of blubber on mud and layer on layer of wailing and baying, burbling and bleating.

I can hear the plangent cry of a curlew too, though it's easy to overlook the bird life when the seals are so absorbing. On any other day, in any other location, I'd have been riveted by the sight of hundreds of foraging shelducks and Brent geese. Starlings are mobbing the sea buckthorn for its flame-orange berries while the magpies weaving among the seals are rather more partial to fresh placenta.

'Those magpies – horrible things – they'll scavenge owt.'

Now that the RAF training flights are over, my fellow visitors, hunched and hooded against the cold, have refocused on the seals, often addressing the pups directly. 'Hello my beauty, haven't you got lovely fur.' 'That were a sneeze and a half. Bless you!' 'Come on, you daft apeth – look at me so's I can take your photo.' The sneezer is sporting an atrophied stub of umbilical cord so must be about four days old. When a slightly older pup, white fur muddied from wriggling and stretching, strays too close, the sneezer's mother grabs it by the hind flippers and gives it a shake, prompting an 'Oo, you nasty bugger' from the man standing next to me.

The very pregnant cow is still sleeping so I shift my attention to the exploits of some of the bulls. According to one of the info boards attached to the fence, both dominant and transient bulls are active here at Donna Nook, just as I observed when visiting Ramsey. The former are physically and sexually mature and capable of defending a group of cows while spending between six and over fifty days ashore. The latter – younger, sexually mature but physically less imposing – haul themselves hormonally around the colony, seeking opportunities to mate and frequently getting banished by their elders.

While it's natural to assume from these behavioural observations that a relatively small number of beachmasters father most of the pups, DNA data acquired from two Scottish colonies over a period of more than a decade suggests that this is not necessarily the case. The majority of the sampled pups were found to have been fathered not by the obviously dominant bulls but by a broad collection of other males who all seemed to enjoy a comparable rate of paternity success. One possible explanation for the disparity between the visual proof of beachmastery and the genetic data is that underwater matings may be more common than first imagined, with a goodly number occurring beyond the terrestrial edge of the breeding site.

It's not long before I spot what must be a transient male lurking in one of the marsh's murky creeks next to a sign advising him not to 'touch any military debris. It may explode and kill you.' Tangling with a dominant bull should perhaps also carry a warning – as soon as the younger seal emerges from the creek and tries to sneak towards a cow resting with her moulting pup, an older male, splashing mud and flashing teeth, crashes over to evict him. Other scuffles are characterised by bulls' kinetic innovations, sliding and aquaplaning being much speedier than their usual cumbersome thumpalong.

I, meanwhile, propel myself towards the epicentre of the viewing area where a Portakabin gift shop and a Lincolnshire Wildlife Trust membership van are sited. There's also a whiteboard displaying the Trust's weekly seal count totals – the current tally is 597 bulls, 1,437 cows and 1,316 pups. The reserve's assistant warden, whose name, Lizzie Lemon, could have been stolen from a character in a children's book, is presiding over this area. Seeing as she's having to check on and manage her volunteers, help out in the gift shop, litter-pick and cheerily greet as many visitors as possible, I'm grateful that she can find even a few minutes in which to chat with me. Our exchange is mostly about the sub-zero temperature – she's relying on hand warmers tucked inside two pairs of gloves today – and the enviable insulating blubber of seals. She refers to the RAF flights too and insists that the seals are completely unaffected by them, just as we would be by train noise if we were living next to a railway line. She also recommends that if I want to learn more about the year-on-year increase in pup numbers as well as the challenges of the associated increase in visitors, I should speak with senior warden Rob Lidstone-Scott a little further along the viewing path.

The arrival of a boisterous primary school group puts paid to any subsequent conversation. A nearby cow, alarmed by the kids' shouts and screams, rugby-tackles her pup, intending to protect it but half-squashing it instead. One of the schoolteachers, intent on linking the field trip to the national curriculum, quizzes Lizzie about including cows and bulls on pie charts and graphs.

I reckon it might be an idea to ditch the maths and do some basic biology instead as, just metres inside the fence, some spectacular attempts at mating are happening. To the left, a bull, all mouth gapes and posturing, approaches a cow who signals her aversion by rapid flipper-flapping and howling. He ignores this and hauls himself on top of her at right angles, pinning her down. Though he's considerably larger and heavier, she manages to struggle out from under him while a second cow, fearing disruption to her pup, charges towards him with a back-of-the-throat growl. Outnumbered, he reluctantly retreats, trailing ropes of drool.

Meanwhile, in the shallow channel immediately ahead, two substantial bulls are pursuing the same cow but when they break off the chase to grapple with each other instead, a third bull, a transient perhaps, grabs his chance and steals in.

'Well, you see life here all right,' says a man in a tweed cap.

The bull moves behind her, grips her with his front flipper and bites the back of her neck, pulling the skin taut till it looks like it'll rip. After a few ill-targeted prods with his penis, he finds the right entry point and they lie there spooning for fifteen motionless minutes. Then, with a subdued howl, she breaks away while he rolls over, exhausted but still erect, in the creek.

Three teenage lads in woolly hats and shorts have been following the action, two of them with thrusts of laughter and one with a furrowed brow. 'Guys, I got a serious question,' he says. 'You know when we went skinny-dipping at Skeggy and we were all shrinky-dinky? Well, that seal's in the water so how come it's not gone shrinky-dinky for him?'

Leaving them to contemplate this critical issue, I wander on in search of Rob. While earlier, I was mostly seeing newborns, I'm now predominantly spotting pups who are at least ten days old and starting to moult their birth fur. One particularly plump pup who's grown all his dappled, adult coat but for a few patches of white still tufting his back is suckling from his blubber-diminished mother. I'm close enough to witness the vigour of it, the gulps rippling through his body, and when he switches teats, I see they're no longer girded by fur – the past couple of weeks of feeding have left her with completely bare skin.

A short pause in the flow of visitors gives a lithe, chestnut-coloured creature the opportunity to dart across the path. Though I get only the most fleeting of glimpses, I note that its tail's tipped with black. Stoat, then, not weasel – and too large, surely, to be preyed on by the male kestrel that's looming over the dunes. I watch him do his fan-tailed hover for a while, wondering if his Northern folk-name of wind-fucker was ever in use here in Lincolnshire.

Hovers of conversation reach me as soon as I walk on again too. A man blithely passing comment on a cow's intimate anatomy ('Her nipples aren't up to much. Nowt to get your chops on'). A woman blowing kisses ('Hey, little baby!'). Widespread hilarity at the scratching, the snoring, the farts. A dayglo-waistcoated volunteer warden spouting the official line about the MOD ('The bombing doesn't bother the seals a bit. And they always check from the control tower to see if there are any on the sandbank first').

I eventually find Rob coning off a short section of the path to narrow it and allow for more of a gap between the procession of humans and a rather unsettled cow on the seals' side of the fence. He's attaching a handwritten sign there too. 'I'm protecting my pup + am a bit grumpy! Please give me a bit of space.'

We give ourselves a bit of space for our chat by walking to the end of the path where the public access runs out and all visitors must turn round and walk back to the car park. Though he's muffled up for warmth, his glasses and beard, which contains all the salt-and-pepper variations of a cow seal's coat, are still visible.

'I first came here one Boxing Day with my parents in the seventies,' he tells me. 'I brought my new toy guns with me and lost one in the dunes – when I applied to be the warden here, I said I wanted to find it!' Whether it was down to this quip or his experience of working with Antarctic fur seals and albatrosses on South Georgia, he got the job. 'When I first started, I couldn't believe how it had changed since the seventies – back then, there was just a handful of seals on the outer sands. And it's kept growing too – this year, we'll probably end up with around two thousand pups.'

I've come across various theories about, but no definitive explanation for, the growth of Donna Nook, and Rob and I briefly touch on a few of these. Population studies have demonstrated that, in recent decades, there's been a decline in the rate of increase, then something of a levelling off of births in the Hebridean and Orkney colonies, so it might be supposed that the concurrent expansion of Donna Nook and other North Sea sites indicates that adult females have been moving south and east. However, research into genetic differences, and population modelling work, have revealed that there's a degree of reproductive segregation between those seals who breed in the Hebrides, those who breed on Orkney and those who favour the North Sea sites. Grey seals travel vast distances to forage but most, it seems, return to their discrete region to breed. There does, however, appear to be some redistribution of breeding females between sites within the wider North Sea region. As with the development of the Isle of May colony, it's also possible that the long-term disturbance caused by the Farnes culling programme could have forced pregnant cows to relocate to Donna Nook. There's the broader European grey seal population and movements between here and the islands and coasts of Denmark, Germany and the Netherlands to consider too. Tagging and tracking of small numbers of Donna Nook seals over the past few decades, plus limited photo ID work, have confirmed that such movements take place but it's thought that the breeding populations in some of those continental locations is being bolstered by an immigration of seals from the UK's North Sea colonies rather than the reverse.

Whatever the reasons for the increase in pup numbers here, there's been a correspondingly significant jump in human numbers. In 1997, at around the time when Rob started wardening, there were approximately four hundred pups and eighteen thousand visitors. Now, as many as seventy thousand folk come to view the seals each breeding season. 'We were on BBC Two on *Nature's Calendar* and that brought in thousands of extra people,' he says. '*Autumnwatch* and *Coast* keep getting in touch but we won't let them film here as we just couldn't cope with any more.'

For many years, Rob had to struggle on here alone but thankfully, in 2012, 'lottery money paid for Lizzie'. One of her many responsibilities is to manage the volunteer wardens who are posted at intervals along the fence to answer visitors' questions and minimise disturbance to the seals, and whose assistance is especially welcome at weekends. 'We had over five thousand people here last Saturday – though there's always a keen core, there are some who'd rather be at Alton Towers and they're not keyed in at all.' Coming here is like a zoo experience for many, he says – the proximity of the animals and the presence of the fence create the impression that the seals aren't wild, or even that they're farmed. One visitor, gesturing towards an afterbirth, asked if they get fed joints of lamb, 'but I've also seen people in tears because they're so moved'.

I tell him how fascinating I've found it in the few hours that I've been here to listen to people's responses to the seals. Meeting in such abundant numbers at our respective species' boundaries – an extremely rare occurrence in these denatured times – appears to be inspiring all manner of reactions from maternal tenderness to outright objectification. And it strikes me that the laughter at the snoring and farting is displaced embarrassment at our own messy and recalcitrant bodies, a subconscious and, for some, uncomfortable realisation that we too are animal, that these seals with their urges to feed, to compete, to breed, are our kin.

There's a brief break in our conversation as Rob nips back along the path to check if people are observing the instruction to give the agitated cow a wide berth. While he's gone, I watch a moulted pudge of a pup roll over to the fence and start mouthing the wire, expanding her exploration of her environment. No nearby cow is keeping tabs on her so I'm assuming the suckling period's ended and her mother's returned to the sea. After a few post-weaning weeks of fasting, hunger will likewise needle her to leave the colony for the first time.

'The MOD paid for that second fence,' says Rob when he returns and spots the pup's mouthing. 'We needed it because when there was only one fence, people could reach through it and touch the pups.' Even so, a double fence doesn't always make for an entirely watertight deterrent. 'One day, I found a couple of Norwegians in the gap between the two.'

'Overall, what's the relationship like between the Wildlife Trust and the MOD?' I ask, still not able to get my head around the fact that a seal nursery and an active military range can comfortably co-exist.

'Very good,' he says at once, arguing that sharing the reserve in this way has actually been beneficial to the seals. Since the range is closed to the public every weekday, people can only access the unfenced, non-wardened outer colony on the sands beyond the marsh at weekends. 'In other words, the seals have their own private air force!' Groups of photographers, often on organised tours from mainland Europe, are particularly problematic and Rob has to do a dawn patrol every Saturday and Sunday to try to prevent them from traipsing all the way out there and crowding round the pups. A few years back, visitors to the outer colony created so much disturbance and caused so many cows to abandon their young that pup mortality more than tripled.

A feeble beam of wintry sun is lighting up a section of the distant stretch of sand on which there's a series of MOD targets, fluorescent orange squares on a blue background. I can make out the humped, dark shapes of bull seals too and, having reached for my binoculars for the first time today, notice one raising himself with his front flippers as if executing a press-up, then letting his chest and belly thump back down against the ground. This behaviour, repeated several times in succession, must be the threat signal known as body slapping. It's unique to the flat, wet sand of Donna Nook and a couple of other east coast breeding sites and certainly wouldn't translate well to the rocky beaches of Wales and the South West. Since it was first observed in the 1990s, body slapping's become an integral part of the ritualised displays of bulls in this area, accompanying their mouth-agape showboating, and sometimes escalating into full-blown brawls.

Rob reveals that if we weren't so far away, we'd hear the smack of body-bulk on sand and also feel the vibration it creates, adding that researchers have spent time here making seismic measurements of the slaps to assess whether the magnitude of the vibrations is a reliable indicator of a bull's size and capacity to win fights. Though non-vocal, vibrational communication is performed by thousands of insects, it's much rarer among mammals, with only an estimated thirty-two species exhibiting such behaviour. It's intriguing to think that the environmental conditions here at Donna Nook and a few neighbouring colonies have led to the evolution of a completely new practice, witnessed nowhere else in grey seals' global range, and probably used by males to evaluate their potential opponents' mass and strength prior to deciding to enter into a tussle.

Raising my gaze above the sandy expanse and body-slapping bull to the horizon, I count a total of eight large tankers, bound perhaps for Immingham just up the coast. Earlier, from a volunteer warden, I heard about Exercise Grey Seal, a two-day operation involving multiple agencies, which tested the National Contingency Plan for Marine Pollution from Shipping and Offshore Installations. The imagined scenario was a collision between a tanker and a passenger ferry thirty nautical miles off the Lincolnshire coast. One of Rob's Wildlife Trust colleagues was deployed to locate mock oiled seabirds in the coastal marshes with the added incentive of a chocolate bar secreted in each spot. The impact of an oil spill on a large seal breeding colony would, of course, be devastating – in *Waterfalls of Stars*, Rosanne Alexander offers a distressing account of pups' suffering on the beaches of Skomer as a result of the 1978 *Christos Bitas* spill while other oil pollution incidents from the 1940s onwards have seen seals stricken with eye problems, respiratory issues, gastrointestinal bleeding, multiple organ damage and contamination of prey.

Underwater noise from marine vessels could also have a detrimental effect as grey seals, especially in the congested North Sea, can't help but extensively frequent designated shipping lanes. In addition to this high spatial overlap, seals' hearing range overlaps with the prevalent, low-frequency range of noise emitted by commercial ships. Despite the fact that they're able to close their external ear canals when they're diving, they've evolved to hear more acutely and detect the direction of sounds more accurately when they're beneath the sea than on land. It could be that the extra-sensitive hearing of weaned pups, particularly when they first disperse from the breeding colony, is most vulnerable to harm from shipping noise, and the extent to which they might be likely to suffer temporary, or even permanent, auditory impairment deserves further investigation. I think of Dad and how dementia's increased his sensitivity to many everyday sounds – as much as he loves Hooper, for example, an unexpected bark can trigger a wince and a pressing of both hands to his ears – and worry that the sea's on its way to becoming similarly overwhelming to seals.

Though Rob needs to return to his wardening duties[1] and I need to return to the hugely pregnant seal near the start of the viewing area to see if she's given birth yet, we spend a final few minutes chatting about one other long-term hazard facing the Donna Nook colony – anthropogenic climate change. The pups here don't have to deal with flooded birth caves or wave-slammed, cliff-backed beaches like those in Pembrokeshire but the threat of rising sea levels and storm surges is ever present along the east coast. In December 2013, Rob tells me, at peak pupping time, a deep low pressure system coupled with strong northerly winds added metres to the already high astronomical tide with the result that water surged well beyond the double fence and footpath. Fortunately, a couple of hours before the high tide was due, staff cut gaps in the fences and opened the gates that were already installed to allow the seals through. They scattered far and wide – one bull was even found inland on his way to North Somercotes – and as many pups as possible who'd been separated from their mothers were transported to nearby Mablethorpe Seal Sanctuary.

'Where are you going, madam?' Rob breaks off from his storm surge narrative to address a cow who's ploughing through pairs of settled mothers and pups to escape an amorous bull. A woman in a woolly poncho who's reached the path's end looks startled, assuming that Rob is addressing her. The man with whom she's walking has one of her hands clutched deep in his duffle coat pocket. They pause at the final section of fence, and he looks out at the coupling, suckling sprawl of a colony with a frown.

'So this is the breeding season is it?'

For the rest of my first day at Donna Nook and for several days thereafter, I wait at the fence next to the very pregnant cow, warmed by numerous cups of tea from the Tasty Treats van. White clumps of fur accumulate in the dunes from pups who are moving through moult. Umbilical stumps dry from red to black and my prevailing anxiety similarly seems to shrivel, a benefit of being absorbed in such a long and concentrated period of seal watching. Hundreds of visitors drift along the path muttering about the cold. And still the expectant cow fails to deliver.

I continue to be astounded by this site and confounded by its complexities – by the fact that the seals have habituated so thoroughly to human presence, and by the ownership role of the MOD. Rob's argument that disturbance to the outer colony is minimal thanks to the military range is a persuasive one, yet the fear and disorientation of some of the pups when I watched the fighter jets roar over was palpable. I suppose, over time, they might grow used to it, just as I desensitised Hooper to potentially frightening noises when he was a puppy by playing a CD of thunder and hoover, fireworks and motorbikes. As a result, he's never been stressed by Bonfire Night but calmly ambles in the garden, sniffing the burning air.

The collision of nature and munitions at Donna Nook also makes me think of Dad as he experienced a similar internal collision throughout his working life. As a shy boy who attended twelve different schools by the age of fifteen, he sought solace and stability in the natural world and would have loved to pursue a wildlife-related career. Instead, though he loathed chemistry and maths, he did what was expected and followed his father into a government laboratory that tested and evaluated explosives.

'Have you been to Grimsby?' he asks me at the end of each day. 'I was born there, you know.'

'I've been to Donna Nook.'

'Where?'

'Donna Nook. A nature reserve a little way to the south.'

'Never heard of it.'

'I've been watching seals there.'

'Never any seals in Lincolnshire when I was a lad. Where else have you been?'

I tell him I've seen very little of his birth county apart from the pupping colony, tractors pulling trailers laden with cabbages and a sign outside a grocer's in North Somercotes announcing that 'Myers Plum Bread is now available!' The latter prompts a pre-war childhood memory of Christmas at his uncle's farm and I promise to bring several of the traditional Lincolnshire loaves back to Wales for him.

Apart from chatting to Dad and reassuring myself that he's okay, I spend the cold, dark November evenings catching up on reading and discover another discomforting link between grey seals and the military. A 2012 press release from the Smithsonian's National

Zoo in Washington DC announced the death of a 'Cold War veteran', a grey seal named Gunnar. Born and recruited in Iceland, he was trained by the US Navy from the age of six months to perform underwater tasks including using a screwdriver and turning a wheeled valve. The Navy's Marine Mammal Program had already enlisted California sea lions, orcas, pilot whales and several species of dolphin, and grey seals were ultimately also conscripted because of their foreflippers being better able than sea lions' to manipulate different objects. However, they allegedly proved to be more wilful and less compliant, and when the Navy decided to ditch the grey seal arm of its Marine Mammal Program in 1979, Gunnar was transferred from the Naval Oceans Systems Center in San Diego to the National Zoo. The press release was at pains to stress that he survived until the impressive age of thirty-eight, exceeding, by over ten years, the typical lifespan of wild grey seal bulls, though there's no assessment of the quality of that lengthy life spent in captivity. Four years later, a similar press release marked the death of a second grey seal, Selkie, who, thanks to her species' dexterity and diving ability, was likewise trained by the Navy to execute Cold War missions, then discarded. Having been rehomed at the zoo, she was conscripted into its breeding scheme, mated with Gunnar and obliged to birth four pups.

On my last morning at Donna Nook before I head to the next seal site on my itinerary, the temperature's again hovering just above freezing with the easterly wind making it feel some way below. From the hide of my furry hood, I watch a bull crocodiling through one of the creeks with just his eyes and the dome of his head visible. A moulting pup mimics her mother's grumpy lunging though aims at a tiny turnstone instead of a gull. Cameras telephoto a cow mid-howl.

I spot a dead newborn on the marsh too, still shrink-wrapped in the amnion. Usually the force of being born causes the membrane surrounding the pup to break but on the rare occasions when this fails to happen, the pup suffocates. Of the mother who birthed it, there's no sign.

As I've done every morning, I stop at the fence near the very pregnant cow alongside a volunteer warden and a hopeful gathering of visitors. She's sleeping on her side, planetary belly towards us, her fur patterns abstract splashes of paint on a stretched canvas. Female seals generally go into labour within a day or so of coming in from the sea so her tardiness isn't typical. Yet, keen though I am to witness a birth, I'm perversely enjoying this exception to the textbook, not to mention attaining a level of patience that I hope I'll be able to replicate in all my interchanges with Dad.

'Any time now, I reckon.'

'Nah, could be hours yet.'

'Someone give her a curry – that'll help.'

We react to every twitch and fidget. A ripple of her belly, indicative of a foetal wriggle, triggers a ripple of anticipation along the fence. She wakes and opens her back flippers, stretching the webbing, then curls them up again like dormice furled in hibernation.

'Take no notice of us, babe. Have another nap if you want one.'

Twin girls, pink-cheeked under matching pink hats, squeeze in next to me. Both are clutching a cuddly seal from the gift shop and both hold them up to the fence so their plastic eyes can watch the pregnant cow too.

Much to their mother's irritation, their brother's not at all interested. 'No, Jacob, you're watching this seal. You can play on your iPad later.'

The cow rolls onto her back, stomach like a tumulus, but, unable to settle, lurches up again and lumbers towards the creek.

'Come on, lass,' urges the volunteer warden. 'Come on.'

Restlessness is usually a sign that labour's looming but, after a brief wallow, she struggles up onto the mud again and slumps back into sleep.

Many of the watching women murmur in sympathy at her efforts to get comfy and start comparing their own labour tales. 'We wanted a water birth.' 'Not me. I wanted drugs.' 'My waters broke in Asda.'

While, on one level, it feels a bit invasive for so many lenses and gazes to be trained on her personal pre-birth behaviour, there's something unexpectedly touching about it too. Occasionally, she drags herself back out of sleep, raises her head and looks at us with indifference – an expression, quite without curiosity and quite without fear, that I've never seen in a seal before, a result of habituation, I suppose – and with an inevitable absence of awareness of how we're collectively willing her on.

'Stop it, Jacob. Put your coat back on – it's freezing. That seal's got her eye on you – she knows you're being naughty.'

Jacob pulls a face, flings his coat onto the path and starts to trot off.

'Come here! Now, I said! Or that seal's coming after you.'

Though many give up the vigil due to the cold or boredom or hunger, I spend my last Donna Nook hour at the fence with a dedicated core audience of ten. I've learnt so much here about the complex grammar of seal–human interaction and my pupping colony behaviour vocabulary has increased exponentially too.

And this pregnant cow, more than any other seal I've encountered, has taught me how to conjugate the verb 'to wait'.

20

The Count

There's no mistaking which conservation organisation owns and manages the next seal breeding site I visit a few hours south of Donna Nook. As is clear from the waterproofs, fleeces, beanies, baseball caps and badges of the two rangers and one volunteer with whom I've just met up, not to mention the oak leaf-branded Land Rover into which I'm about to squeeze, it's again the turn of the National Trust.

Blakeney Point is a four-mile shingle spit that reaches into the North Sea like the arm of a string puppeteer, with the tapering shape of its western end like the downward curve of his fingers. It's a cold, grey day and I've come to the car park at Cley beach on Norfolk's north coast, having managed to get myself invited along on one of the Trust's twice-weekly pup counts.

Ajay Tegala, the profoundly polite ranger with whom I've been liaising, sandwiches himself into the Land Rover's front seat between me and Sally, the retired teacher volunteer, who'll be driving. In the back, among piles of other gear bearing the NT logo, is the second ranger, Carl, plus a man who's been booked to test the alarm system of Blakeney Point's visitor centre. Cramped though his position is, he's drumming his hands on his thighs with anticipation. 'We gonna see some seals, then?'

The Count

Considering that Blakeney Point supports the largest breeding colony of grey seals in England, there's a strong likelihood that some will indeed be seen. Like the Farne Islands and Donna Nook, it's another east coast location that has experienced a steep rise in pup numbers in recent decades from just two in 1988 to twenty-five in 2001 to the first season when over two thousand births were recorded in 2014.

Sally accesses the shingle bank a tad cautiously, failing to get purchase until Ajay advises her to rev. Our eventual bumping and ploughing through the pebbles is punctuated by his occasional gear changes on her behalf and the alarm guy's shout of 'Seal!' every time he spies one bottling to our right in the water.

When we're about a mile from the end of the spit, we drive past a National Trust 'No Further Please' sign, as well as what looks like a dead, washed-up whitecoat. Sally shifts from the pebbles onto a sandy track with some relief until another hazard presents itself. Though we're bypassing the core of the pupping colony, a number of bull seals – probably all transients who haven't managed to maintain access to a group of cows – have strayed to the periphery and are lolling on, or adjacent to, the route we have to follow. As soon as they become aware of our approaching vehicle, most galumph off into the dunes though Carl needs to clamber out of the Land Rover to clap and wave his NT fleece at one especially slumberous individual to encourage him to move to a safer position. By the time we draw up at the visitor centre, the alarm guy's eyes are as wide as the Norfolk sky. 'This is the most amazing job I've ever been on!'

I'm thankful, given the appealingly remote feel of the location, that the visitor centre doesn't include any multi-media displays or a jam-and-chutney-stuffed gift shop but is on a much more modest scale. It's housed in a former lifeboat station, painted bright blue, which doubles as Ajay's summer accommodation. He conveys, in his softly spoken way, that as well as having a deep interest in the coastal flora, he's responsible for monitoring the Point's significant population of nesting birds. Sandwich terns – the UK's biggest colony – and the amber-listed little tern are among the species that raise chicks here during the summer months.

While the alarm guy dejectedly starts work inside the visitor centre, Ajay, Carl, Sally and I set off on our pup count, briefly joining a boardwalk before ducking under a flimsy barrier to access the colony. In contrast to Donna Nook's double fence, it consists of nothing more than orange twine tied between a series of rusty stakes. Ajay admits that the relative inaccessibility of Blakeney Point – visitors have to trudge for four miles, often through ankle-deep shingle, to reach the seals – makes his job much easier than that of the rangers in some of the other east coast colonies. 'Our busiest time will be about a month from now, right after Christmas – lots of people come on holiday to North Norfolk plus there was a tradition of local families coming here to walk off their Christmas lunch long before the seals started using it for pupping. On New Year's Day, we can get around eighty people so that's when we need a volunteer presence.' He only has about twenty-five volunteers to call on, however, further proof that, in comparison with busier seal sites, their services are not in high demand.

While Ajay's been talking, *eau de* bull seal has surged into my nostrils and we emerge from the cleavage between two dunes into an area of saltmarsh. Here, there's a gaggle of moulting pups and vigilant cows plus the source of the smell – two sizeable bulls trying to out-body slap each other, and sliding through the mud to get close and square off.

'I wonder if they enjoy doing that,' says Sally.

'I imagine they realise it's useful.' Carl sounds a bit disapproving of her drift into anthropomorphism.

For ease of counting, the colony's divided into eight sections and it's decided that we'll split up and share them between us. Carl and Sally will work in tandem while I'll accompany Ajay. Before we head off, he runs over what we should be looking for: we only need to record pups aged three days and under because those who are older have already been included in previous counts.

In other words, we're seeking small, saggy-skinned bodies with well-defined necks not yet bloated with blubber; umbilical cords that are pink or red rather than brown and shrivelled; and fur that's probably still yellowish with amniotic fluid like the golden crust on clotted cream.

The first section on which Ajay and I must focus is ten minutes' walk away and I'm glad that I don't need to start recording pup numbers till we get there. As we creep through the heart of the colony, the surround-sounds, smells and sights outpace my brain's capacity to process them and any counting I'd do would be erratic in the extreme. To our immediate left, a cow glances from her pup to us, hissing a throaty caution. From the right, a lunge, a feigned attack, then quick-shunt back to her whitecoat. Ahead, an older pup, one of hundreds currently in view, rests in a nest of flattened marram grass and shed fur. His nostrils flare and he mimics the cows' hisses, giving a guttural utterance that's more like a snore than a warning. Sanderlings scamper round the afterbirths.

All the pupping colony activity that I've witnessed at Aber Felin from the cliff top and at Donna Nook from behind a double fence is unfolding just metres from my feet. It feels wondrous to be weaving through this agitation of breeders yet also quite transgressive. Having devoted so much time, over the course of my journey, to talking, and thinking about, the importance of giving seals space and keeping an appropriate distance, I have to keep reminding myself that this visit is permitted, that we're gathering information which will contribute, albeit in a very small way, to future scientific analysis of the species' health and resilience.

Never walk between a mother and her pup. Never block a seal's route to the sea. A seal bite can be serious – the bacteria in its mouth can cause 'seal finger', an infection that, without specialised antibiotics, leads to swelling, inflammation, joint damage and, in the worst-case scenario, amputation. While my years of seal watching are undoubtedly helping me read behaviour and anticipate how those around us might react, there's still a degree of unpredictability and I'm not going to downplay the danger. I recall Kimberley Bennett's comment about some of the Isle of May cows wanting to attack her kneecaps when she's ferrying blubber samples through the colony too so keep as close as possible to Ajay as he picks the route that'll cause the least disruption.

We inch our way past an indifference of bulls – they barely seem to register our presence, focused as they are on the fertile cows and each other's next move – then undulate over dunes stacked with weaned and moulted pups like barrels in a brewery. And the whole time that I'm sidestepping and sidling and absorbing, Ajay's whispering details both about his work and his friendship with Bee and Ed, the wardens on Skomer. Ed used to be the Blakeney Point ranger till Ajay took over and when he visited them in Pembrokeshire, their abseiling down the cliffs offered an arresting contrast to the way in which the pups are counted here.

We've reached the start of our first section, a low dune ridge on the sea side of the colony with a beach to our right. Ajay hands me a clicker counter and, while I faff around, working out how to accommodate it alongside my camera, binoculars, notebook and digital recorder, he offers a last-minute tip. 'Don't count ahead as you'll get confused – only count a pup when you're actually walking past it. The tide's a fair way in so be aware that they're quite clustered – it's easier to count when the tide's low and they're more spread out.'

We start walking side by side, Ajay looking into the dunes to the left and me towards the beach, left thumb poised to press the clicker. I soon discover that it's not easy making a quick, definitive assessment of the age of a recently born pup. The just-borns, with the moist, pink remains of the umbilical cord are unmistakeable, as are the more mobile, five-days-plus pups whose cords have sloughed off. But is the one with the umbilicus that's red with tinges of brown, dry and semi-shrivelled, three days old or four? Should I click the counter or assume it's already been included in Ajay's last visit?

'How are you doing?' he asks kindly, picking up on my dithering.

'I've got a few that are borderline.'

'That's to be expected. So have I. How many are you unsure about?'

'Five.'

'Let's take three of them.'

This is, I guess, about as accurate as it's possible to get when counting at a site where there's no dye-marking of newborns. Human error and inconsistency are bound to play a part especially if different people are collecting the data and interpreting the guidelines as to what defines a three- or four-day-old pup in a slightly different way. Better to settle for the minor imperfections of this method, though, than to move through the colony twice a week, spray-painting the new arrivals, which would cause a far more serious degree of disturbance.

Accepting that my monitoring technique will never be 100 per cent accurate helps me start to relax into a more comfortable counting rhythm, and I only pause at the sight of a cow with strange orange belly stains, which give the impression that she's rusting. She's shifted onto her side to issue the classic moaning, flipper-flapping rebuke to an over-attentive bull, so the stains are fully exposed. I've seen new mothers' bellies and flanks smeared red from birth blood and inadvertent rolling on the placenta but this cow's pup is older than newborn and the colour of this stain is more like a dodgy fake tan.

'Ah!' says Ajay, following the direction of my gaze. 'An Essex seal.'

'A what?'

'A seal that must have been hauling out around the coast of Essex – the mud there contains iron oxide. We see quite a few of them here.'

We carry on westwards towards the end of the Point, the shape and position of which are continually shifting, thanks to the transporting and depositing of sediment by the waves. As has been the case with views from many of the seal sites I've visited, ranks of wind turbines reiterate themselves along the horizon. Ajay points out the route that local boat operators take from the harbour of Morston Quay when bringing visitors to the Point to view the seals. At peak times, several different companies offer a range of trips per day and they're especially popular while pupping's underway – indeed, it sounds like they approach Farnes-level intensity. However, Ajay argues that the fact that people have the opportunity to view the seals in one part of the colony by boat means that far fewer are tempted to slog along the shingle ridge and access the pups on foot so the degree of disturbance is much less than it otherwise might be.

'The boat trips are important to the local economy too,' Ajay adds, 'especially at this time of year when more people are out of work and— Oh, crumbs!'

His exclamation's a bit quaint for the disconcerting sight that greets us as we walk back towards the dunes from the end of the spit in readiness for our next phase of counting. A great black-backed gull is limping across the shingle in front of us, dragging its damaged, almost severed, right wing behind it.

'Close encounter with a bull perhaps?' says Ajay.

'Or a cow,' I suggest, 'if the gull was trying to grab the placenta and got too close to her pup.'

'Ah yes, that seems more likely.'

We watch it struggle towards the edge of the dunes, a shadow of its healthy, predatory self. I'm so used to seeing this species, all strut and bluster at a birth site, and it's a sobering reminder of the strength of a cow's urge to protect her pup.

'It'll never fly again, that's for certain,' says Ajay. 'I suppose it may find enough to scavenge on for now if a fox doesn't get it first. But when the seals go, all there'll be to eat is the odd dead hare.'

After another few hours of counting, we're ready to scavenge through the contents of our rucksacks for lunch. By now drenched in the stench of hormonal bull and fresh placenta – though I have yet to witness a birth – we sit on a mound of sand, with a view of both receding sea and expanding colony. Ahead, a hefty male is flinging his front flipper across a cow's back and biting her neck in an attempt at mating. To our left, no more than ten metres away, another bull is resting post-territorial brawl, blood and marram grass stalks clotted in the folds of fur at his throat. And below the wails of hungry pups and the howls of angry cows, I can hear the suck and swallow of newborns feeding.

Between precise bites of his bap, Ajay speculates whether today could be the day that the record's broken for the most number of pups ever born on Blakeney Point, as it was almost within grasp after the last count three days ago. 'We always do a proactive press release at the start of the pupping season, often asking if it'll be another record-breaking year, so there'll hopefully be lots of people waiting for news.'

'Don't those kinds of press releases run the risk of attracting unmanageable numbers of visitors though?'

He sighs. 'It's difficult. As the National Trust, we do need to be seen to be encouraging visitors – because of the education side of things too. It's in our remit.'

'What about wildlife filmmakers?' I ask, remembering that Rob and Lizzie don't allow any at Donna Nook for fear that the site will be even more overwhelmed.

'We get lots of requests but we can pick and choose the ones we want to be here. *Winterwatch* always gets the Trust good publicity. A few years ago, they did some night filming and though visitor numbers went right up on the weekend after the show went out, it was towards the end of January and the seals had pretty much all gone by then!' This triggers a bout of robust laughter that's quite at odds with his genteel demeanour.

While he phones Carl to find out how he and Sally are progressing with their count and arrange a time for reconvening, I warm up with tea from my flask and focus on the resting bull. He's subsided from head raised and watchful to eyes closed and chin on sand. When I look back on today, it won't be the anticipation of record-breaking pup numbers that I'll most cherish but this dune time, picnicking in proximity to a sleepy top predator, remaining cautious, of course, yet at the same time feeling deeply at ease.

Though the bull is fight-weary at the moment, he's remained in the heart of the colony, suggesting he's not yet vanquished, neither a young transient on the fringes nor an old one past his prime. I've often wondered what role the older males play at breeding times – those who may be approaching their quarter century, the bull seal equivalent of Dad's twilight years. Do they continue showing up and valiantly grapple for supremacy? Do they reach the point of conceding, in contrast to Dad's reluctance, that their strength is on the wane? At the Cornish Seal Sanctuary, Dan Jarvis didn't convey how peaceably the dominant male baton passed from Flipper to Yule Logs and I'm not aware of any research that examines the interactions of old bulls in the wild either. The challenge of estimating a seal's age perhaps precludes studies of this kind – a bull may look like a battle-scarred veteran but only the removal and sectioning of one of his canine teeth can offer a clearer indication of how old he is. It's possible to count the growth layers, the alternating light and dark rings of dentine, that are believed to be caused by the seals' changing seasonal activities such as fasting during the breeding period.

'Yes. I see. Hmm, yes.' Ajay, whose phone manner – always 'yes', never 'yeah' – is as pleasant and decorous as the rest of him, is listening attentively to Carl. It turns out he's spotted a cow whose neck is trapped in netting – she doesn't appear too encumbered at the moment but the situation will need to be monitored in the weeks ahead. 'We try to free entangled adults,' says Ajay, 'but that's not possible in the middle of a pupping colony.'

'What's your policy as far as rescuing pups goes?'

'Minimal intervention. There's a low mortality rate here, the colony's doing well so it's mostly about reputation management. If a pup's abandoned in the main part of the colony, we don't do anything – no one will know about it. But if it happens on the fringes where people might notice it, we'll do a rescue so they can see the Trust is looking after the pups.'

A pragmatic, if rather disquieting, policy, then, similar to that which Lincolnshire Wildlife Trust follows at Donna Nook, as I heard from one of the volunteers.

It turns out that Sally's driving the alarm guy back to Cley and since Carl has completed their count, he hooks up with Ajay and me again while we finish our last section. We resume our stealthy walk through the dunes with Carl carrying out a one-sided conversation with the seals under his breath. 'Is that all you've got? You've got to do better than that. Must try harder,' he says to a moulted pup who's hissing a weak warning. Then, when we spy a cow, both eyes cloudy and blind, 'Look at that! You're raising a pup! I take my hat off to you!' And he does, indeed, doff his baseball cap with the National Trust logo.

At last, we follow the line of twine and rusty stakes under which we ducked at the start of the count. The seals aren't respecting this fence that defines the edge of their colony either, spilling out onto the boardwalk. We click our counters for the final few times, then head towards our arranged meeting place with Sally. Instead of picking us up at the visitor centre, she's waiting at the No Further Please sign near the point where we turned off the pebbles onto the sandy track in the Land Rover earlier. She's leaning now against the side of the vehicle, her cheeks patterned, thanks to the chill air, with a network of capillaries like tidal channels in a saltmarsh.

Seals are starting to turn even this stretch of the spit into a marginal area of their colony. A whitecoat's huddled under a mobile bird-hide-cum-info-shack, directly below a notice advising What To Do If You Find A Lone Pup. A bull with tiny pebbles embedded in a wound on his head is skulking around a resistant cow. Her close-to-moulted pup is sheltering in a depression in the shingle, one of the Land Rover ruts. It appears that the pup is crying – in order to keep the surface of their eyes moist, many mammals secrete tears that are carried away by the nasolacrimal duct but as seals lack this anatomical feature, their tears, protecting against salty seawater and sand, overflow and trickle down their faces instead. This semblance of weeping is surely another reason why humans have such a strong emotional affinity for seals.

After a brief discussion, Ajay and Carl decide that the No Further Please sign needs to be shifted a few hundred metres back towards Cley because the seals have moved beyond its present position and will be at risk of disturbance if any visitors come trekking along the spit. While they dig it out of the shingle, stagger along with it, then dig two new holes so it can be re-sited, I opt to check out the dead pup that I glimpsed from the Land Rover on our journey here. I'm expecting to find an underweight body, a whitecoat that had lost contact with its mother and starved to death. Instead, however, I'm greeted by the body of a larger pup with a very distinctive injury – a smooth-edged wound, starting at the head and corkscrewing around the body, with a spiral strip of skin and blubber detached from the underlying tissue.

Had I not met with Ailsa Hall at the Sea Mammal Research Unit in St Andrews earlier this year, such a discovery would have left me flummoxed. She explained that isolated spiral injury strandings started to be reported from the 1980s, but after more concentrated clusters of corkscrew carcasses of both adult common seals and grey seal pups were washed up along this north coast of Norfolk, Ireland's Strangford Lough and Scotland's east coast through and beyond the first decade of the twenty-first century, SMRU was commissioned to investigate. The scientists hypothesised that the injuries were consistent with the animals being dragged through boats' ducted propellers and experiments involving wax and silicone miniatures of seals being fed into a scale model of such a mechanism seemed to validate their argument. Meanwhile, corkscrew corpses had concurrently been appearing in Germany and, most profusely, on Canada's Sable Island where attacks by Greenland sharks were mooted as the likeliest explanation for all the deaths. An episode in 2014, however, led to a rather more startling revelation – during the Isle of May's breeding season, a grey seal bull was observed to be killing and consuming grey seal pups.

Researchers video-recorded an adult male on the island catching a weaner, dragging it to a pool and forcing its head under the water. He then proceeded to bite and tear the pup's skin and blubber layer, causing the characteristic corkscrew injury. Over the next week, the same male was filmed killing, and eating a proportion of the blubber of, four more weaners, and a further nine carcasses, the majority of which bore similar injuries, were found in the same area.

It's not yet known how typical such episodes are within the grey seal population. One theory is that an individual rogue bull might be responsible for multiple deaths across a particular region – the species travels widely, after all, and the Isle of May culprit was later traced to Germany. It's probable, however, that the behaviour – which may enable fasting bulls to remain for longer in the breeding colony and have more mating opportunities since pup blubber is highly calorific – is more widespread and being passed on via social learning, as happened with body slapping.

It wasn't especially comfortable to discover that the adult male of the species with which I've developed such a strong connection displays infanticidal tendencies. Even Ailsa, a supposedly dispassionate scientist, described the Isle of May video footage as 'quite shocking'. But, she added, if it turns out to be 'natural evolutionary behaviour, then that's the way of the world'.

Back in the Land Rover, with Ajay in the cluttered rear this time and Carl squished between Sally and me, I share the corkscrew carcass photos I've just taken.

'I'd say we've had perhaps a hundred corkscrew seals wash up here in the past five or six years,' says Ajay, adding that, before the SMRU evidence emerged, the propellers of boats associated with nearby wind farm construction were thought to be causing the injuries.

One of the things that I find most puzzling, I tell him, is that in all the time I've spent observing grey seal breeding sites, I've only ever seen bulls completely ignoring the pups.

Ajay confirms that this has been his experience too. 'On every count I've done, they've always just been focused on the cows.'

He falls silent for the remainder of the journey back to the car park at Cley beach so he can tot up pup numbers, adding our, and Sally and Carl's, newborn tally to the running total from three days ago. The tide has noticeably fallen, with sand now fringing the shingle ridge along which we're driving and our view of bottling seals more distant. While I can't deny it would be exciting to know that the record for the number of pups born on Blakeney Point was broken on the day of my visit, I'm also feeling a little uneasy about it. I keep thinking of a comment that Sue Sayer made while she was showing me photos from her ID catalogue and simultaneously talking about Cornwall Seal Group Research Trust's crusade to reduce disturbance. 'On the east coast,' she said with a sigh, 'it's just all about numbers.'

Having now been to several east coast colonies, I can, to a certain extent, understand her concern. When there's so much hype around bumper numbers, there's a danger that people will assume that the grey seal population is universally booming while the wider context reveals that, elsewhere in the UK, growth has slowed, is levelling, or has already levelled off. There's a risk that the slew of anthropogenic threats that seals are facing, some of which have only emerged relatively recently and the consequences of which have yet to be quantified, will fail to be taken seriously too. As a counter to Sue's comment, however, the seal issues with which I've engaged on Blakeney Point today, and the discussions I've enjoyed with Ajay, have in no way been limited to the numerical.

We bump down from the shingle and draw to a halt, back where we started, on the concrete surface of the car park. Carl and I jump from the front of the Land Rover while Ajay climbs out of the rear, notebook in hand and smiling.[1]

'Two thousand, four hundred and ninety-eight pups,' he says. 'It's the record.'

21

Birth and Rescue

Howls and grunts and wails are bringing depth and texture to the pre-dawn dark. From my dune-top spot, my eyes gradually acclimatise to the lack of light and I manage to make out hordes of hulking forms persistently shunting themselves into new configurations on the beach. Soon, a bloom of peach ahead on the eastern skyline signals the sun's forthcoming arrival and the shapes become identifiably seal. A bull hails daybreak with a wave of body slaps on the sand. A young pup stops suckling and hiccups over and over, its whole body convulsing, while a moulting pup rolls in the foam at the sea's edge with its mother.

It's early December, close to 8 a.m., minus three degrees. I've reached Horsey, the last of the major grey seal pupping colonies on my itinerary: like Blakeney, it's Norfolk-based. As the space-hopper sun bounces up above the horizon, it spotlights the frosted spikes of marram grass among which I'm huddled while the rope strung along the top of the dunes to discourage access to the seals glistens with crystals of ice. In spite of the cold, there's a hint of California about this place with its beach, purfled with surf, extending for miles in both directions, albeit interrupted by a sequence of groynes, some of them built from timber and some from rock.

The low sun is reddening the rust of the groynes' supports, buttering the pups' white coats, brightening the bellies of cows turned sideways to suckle. It's highlighting the neck wounds of fighting bulls too, as well as the blood that's puddled on the beach right below me: thanks to the earlier darkness, I've once again just missed witnessing a birth.

The newborn's fur is a mustard-yellow mix of amniotic fluid and sand. Though it's eager for its first feed, its mother has moved a short distance away as she hasn't yet expelled the placenta. Partly exposed between her rear flippers, it's being targeted by a herring gull darting at her from behind. As she swings round to fend the bird off, the afterbirth slides out completely, cuing over twenty gulls, both herring and great black-backed, to swoop in to seize, rip and scuffle. One screeches ownership of the placenta, blood-streaked throat raised to the sky, while another spreads its wings and tries to mantle over it like a raptor that's captured a rabbit.

The colony stretches for several miles along the northeast coast of East Anglia, the part that bulges like a Victorian lady's bustle. The dunes border its whole length, separating the beach from the low-lying inland pasture, villages and waterways, and the wall with which they are reinforced is a literal concrete reminder of the area's vulnerability to flooding. Constructed from 1953 onwards in response to the North Sea storm surge, it's stepped on the beach side, with the different levels offering extra sleeping space for the moulting pups in particular.

Occasional gaps in the sea defences allow public access to the beach at non-breeding times. My viewing spot above the beach is adjacent to Horsey Gap, site of a large car park and a refreshments van prematurely decked in Christmas tinsel. Come 9 a.m., I stride down from the dunes, hoping that movement and tea will bring heat back into toes and torso. A few visitors are starting to arrive, including a springer spaniel who's racing round the car park, barking and skidding on frozen puddles, plus her flustered owner ('Whatcha barking for? You ain't seen the seals yet. Wait till you see the seals.')

The first seal wardens of the day are arriving too, some in yellow hi-vis waistcoats and one supervising duty warden in blue. The latter, for whom I've been told to keep a lookout, is Gemma Walker – before she starts her shift, she's promised to give me an insight into the Horsey colony.

'Sorry – in this cold weather, I've forgotten your name!'

Once I've reminded her, she greets me warmly, face rosy with the outdoors, then leads me back up onto the dunes to the designated seal viewing area. Though she works as a community engagement officer with Norfolk Wildlife Trust, she's here today in her role as a volunteer with Friends of Horsey Seals, the origins of which can be traced to a wardening project, developed from 2002, that aimed to reduce disturbance to the burgeoning colony. Like other breeding sites I've visited along the east coast, pup numbers have appreciably increased here in recent decades from about fifty births in the first years of the twenty-first century to a current annual total of around two thousand. In 2011, Natural England and the Broads Authority, the organisations that had been funding the wardening scheme, withdrew their financial support but existing volunteers established a community group to enable their work to continue. Friends of Horsey Seals subsequently gained charity status in 2016 and now has several hundred wardens on which to draw. While this is a substantial number, there are over five thousand shifts to fill between November and January and seventy thousand visitors to monitor, including more than two thousand on Boxing Day alone. Wardening this site feels like a very challenging undertaking – the seals are as accessible as at Donna Nook, the area needing to be overseen is much more extensive and there's only a rope, rather than a double fence, behind which to encourage visitors to stay.

'At the beginning it was madness,' Gemma admits, tugging her somewhat moth-eaten woolly hat a little further down over her ears. 'You'd have people on the beach with the seals, flying kites, playing rugby. . . Every time I wardened I got shouted at by visitors saying I couldn't stop them going on the beach 'cause they'd been using it for years. Putting the rope up helped a lot – it gives people a clear boundary.'

Not, however, an unbreachable one. Looking down at the beach, I can not only trace the parallel front flipper depressions of numerous seal tracks but a scattering of foot- and pawprints too. Gemma concedes that occasional dog walkers still try to access the beach when the wardens aren't around and mentions the erstwhile tension between a minority of locals and Friends of Horsey Seals – preposterous references to the wardens as 'Little Hitlers', deliberate cutting of the rope, objections to the volume of weekend traffic.

A pheromonal whiff of bull wafts up to us from a male who's approaching a cow in a markedly meek manner. 'Some people think it smells like crude oil,' says Gemma, 'and ask me if there's been a spill somewhere.' After he's rested his head on the cow's back for a few minutes, she shifts her position, makes willing, nuzzling contact and doesn't resist at all when he heaves himself behind her to mate, even though her almost moulted pup's lying right beside her. 'The beach is like a game of chess at pupping time with all the different moves the seals make. And it's like you've got Attenborough in high-definition widescreen, an amazing wildlife spectacle. How many places are there in the world where you can watch large mammals mating and giving birth in close-up? These seals are our megafauna.'

The man and woman who've wandered up the dunes to join us at the viewing area seem rather less enamoured by what they're seeing. 'Look at them going at it first thing in the morning,' says the man. 'Get out of it!'

'Are they a family?' the woman asks.

Gemma discloses that the pup who's currently being raised was conceived a year ago, that the father was probably a different bull who otherwise has no parental input, and that successful males mate with between six and fifteen females each breeding season.

'Fifteen!' A look of horror colonises the woman's face and I'm struck again by how we project our own preoccupations and prejudices onto our encounters with this nonhuman species.

Gemma and I edge away from the couple's disapproval along the dune-top path. As we walk, I spot a number of dead white-coats, one in a shroud of sand, another being watched over by its mother as if she's still expecting it to suckle, a third with empty sockets where its eyes would have been had gulls not pecked them out.

'We used to go onto the beach and bury the dead ones,' says Gemma, 'but we leave it to nature now, even though some parents complain that their kids find it distressing.'

We drop down to the path behind the dunes, as Gemma wants to take me to the second of the viewing areas established by Friends of Horsey Seals, about twenty minutes further on near the next gap in the flood defences. As at Donna Nook, there's a seal quiz to keep children occupied en route, with questions ('What is the largest mammal breeding on British shores?' 'How much can an adult grey seal weigh?') periodically attached to posts alongside the track. A boisterous group of twenty-somethings are shouting out possible answers and high-fiving when they guess correctly.

Having paused to watch a male hen harrier circling over the marshy pasture to our right, we continue to Crinkle Gap where a Friends of Horsey Seals hut and a volunteer warden are stationed on the landward side. There's a donation bucket and leaflets encouraging FoHS annual membership – the charity's only sources of income – plus some robust pig-herding boards that are used for shepherding seals who stray too far into the dune viewing zone back onto the beach. A blackboard displays the numbers recorded during the most recent weekly seal count – 1,700 adults and 1,420 pups.[1]

The warden invites us to dip into a tin of Quality Street but warns that 'the mice have been at them' so we opt to walk on instead. A fingerpost points us in the direction of the Seal Colony Viewing Area – the dunes are steeper here and we follow a straggle of visitors up a flight of wooden steps, with Gemma announcing, as we go, that 1,420 pups is a record number of births for this stage in the season. While this is heartening to hear, the broader context again needs to be appreciated – a higher percentage of pups survive the weeks from birth to independence here on the east coast than on the wave-bashed, cliff-backed beaches of the west but, post-weaning, some 30 per cent may still fail to reach their first birthday.

In this final quarter of an hour before Gemma assumes her duty warden role, we continue to enjoy the pupping activity while she simultaneously fills me in on some of the incidents of disturbance, both wilful and unwitting, to which the Horsey seals have been subjected. She admits that people can be well meaning, doing things with the best of intentions, but that their actions can still have serious consequences. 'They think seals should be in the water all the time like dolphins – we've had them picking pups up and putting them in the sea 'cause they think they're stranded, and of course, they can drown. People bring packets of mackerel and tins of pilchards from the supermarket to feed them too.' She adds that the increased availability and popularity of mobile phones with cameras was a major turning-point, with people venturing ever closer to the seals: most shockingly, a man recently had to be stopped from sitting his toddler on the back of a pup so he could take a photo.

To our left, a woman pauses at the top of the stairs and appraises the view of the colony. 'You'd think they'd flipping freeze to death, just lying around like that,' she declares, though she herself is shivering in just a lightweight denim jacket. Since, in addition to possessing blubber and fur, seals are blessed with the ability to regulate their body temperature via their miraculous vascular network's countercurrent heat exchange system, freezing to death is an unlikely outcome. The veins and arteries in, for example, their flippers are intertwined, enabling the former to capture warmth from the latter and pump blood back into the body's core at a higher temperature.

As if also to disprove the woman's accusation of inactivity, a huge bull whose body-slaps on the sand and rasping exhalations have been ignored, launches himself at a younger male for approaching a cow in the group he's defending. At Donna Nook, the bulls seemed to back off after a short tussle but this fight, teeth ripping into neck, body slamming into body, is intense and protracted. The cow also lashes out with a howl, concerned that the fighters might crush her pup, but, after a brief lull, with the smaller bull slumped and coated in sand like a crumbed sausage, the brawling continues.

The woman in the denim jacket is deeply alarmed by what she's seeing. ''Scuse me!' she calls to Gemma, 'Aren't you going to break that up?'

Battle of Britain Breakfast. Dambuster Breakfast Bap. Prisoner of War Porridge. Haw Haw Hot Buttered Toast. Doodlebug Bangers and Mash. Home Guard Spam Fritters. Winston Churchill BLT.

I'm warming up in the Second World War-themed Poppylands café, a short walk from Horsey Gap. Having eschewed the Franklin D. Roosevelt Americano and the Rita Hayworth Pot of Specialist Tea in favour of a Glenn Miller 'In The Mood' Hot Chocolate, I'm taking some photos of the menu, with its ration-book front cover design, to show to Dad when I head home to Wales in a few days' time. I also need to phone him, having missed making my regular early morning call due to the lack of network coverage in the dunes.

It takes longer than usual for him to answer and, when he eventually does so, he sounds a bit breathless and befuddled. 'This is a funny time for you to ring. What're you phoning again for?'

'We haven't spoken today yet, Dad.'

'Yes we have. Why're you checking up on me?'

'I didn't phone first thing – I couldn't get a mobile signal.'

I hear a few more puffing exhalations, like a speeded-up version of a bull seal's warning. 'Are you okay, Dad? You sound like you're panting.'

'Course I'm okay – why shouldn't I be? I've been up and down the stairs a few times, that's all. I'm trying to sort out the boiler.'

'What's wrong with it?'

'Nothing's wrong with it. Why d'you always think there's a problem?'

Of all the moods on Dad's dementia emotions palette, I find the truculent, slightly paranoid one the most difficult to respond to. *Patience*, I remind myself. *Remember how you waited for the seal to give birth at Donna Nook. And don't take his behaviour personally.* 'Okay, so what is it that you're trying to sort out?'

'The whatchamacallits – the timings. It's turned cold so I want the heating on for longer in the morning.'

I stifle an exasperated sigh. Before I left for Donna Nook, we jointly drew up a list of jobs for which he might need a bit of assistance from Bethan or one of the other carers and changing the heating settings was top of that list. 'It's a really good idea to get some extra warmth in the house,' I say, 'but—'

'Oh yes, bloody wonderful idea. The ruddy thing won't come on at all now.'

I close my eyes and try to picture the control panel on the boiler above the bookcase in my childhood bedroom. It's got a mechanical, rather than a digital, timer, but there are still multiple dials and switches to negotiate and fiddly little tabs, each of which corresponds to a fifteen-minute chunk of time, to push in and pull out. There's no way I'll be able to talk him through the programming process from here.

'Bethan or one of her colleagues will be with you soon. How about asking her to give you a hand?'

'I don't need Bethan to help me.'

'Dad, please—'

'I don't want anyone interfering. I can sort it out myself.'

'You're going to get cold—'

'Then I'll put an extra jumper on. How d'you think we managed in the war?'

Irritation's shortening his breath again. If I continue to badger him, he'll be even less inclined to accept any help. I latch on to his mention of the war instead and start to describe the Poppylands menu. As soon as we've finished our chat, I'll contact the care agency to ask if today's visit can include some assistance with sorting out the heating. By the time Bethan arrives, he'll have hopefully calmed down again and maybe forgotten all about this fractious discussion.

'So, Dad, which do you think I should order? A V for Victory All-Day Vegetarian Breakfast? Or a Women's Land Army Lunch?'

Over the next few days, I fully absorb myself in the dynamics of the pupping colony, just miss witnessing yet more births but grow ever more familiar with the behaviour patterns of the cows, pups and bulls that are visible from the two viewing areas. On my last day before returning to Wales, however, I decide to explore further by walking for several miles along the top of the dunes, following the progression of the colony from its southerly fringes all the way back to Horsey Gap. As pup numbers have increased in recent years, so the colony has expanded as far as the seaside village of Winterton-on-Sea, with Friends of Horsey Seals accordingly expanding their wardening orbit. Gemma indicated that there'd been some local, vocal resistance to the suggestion that sections of the beach should be roped off to protect the seals and though the wardens are doing their best to keep disturbance incidents to a minimum, it sounds like a testing situation.

The Winterton edge of the colony is, I find, populated by a gathering of young bulls, non-breeders who seem to be working on honing their beachmaster skills, play-fighting and body slapping in preparation for future breeding seasons. Then, with the first scattering of cows and pups come the first attendant gulls and humans. A couple of kids creep up on a sleeping, pregnant cow then leap back with a scream when she opens her eyes and registers their presence. A French bulldog strains on her lead towards a newborn, whimpering with excitement (*'Leave it, Princess!'*). A photographer crouched near a chubby weaner urges his girlfriend to 'Help me get its attention', and when she takes off her glove and flaps it, the pup lets out a hiss and hefts itself away. Wardens are dotted on the beach and through the dunes but it's impossible for them to intercept every encounter.[2]

Happily, the further I walk from Winterton's beach car park, the fewer the number of harassing humans and, for the next half hour, the only other person I come across is a man who's anxiously scanning the horizon with his binoculars. 'Where are the sharks?' he says. 'With all this easy prey around, they've got to be out there somewhere.'

This is the first time on my journey that I've happened upon an expression of concern at seals' potential to attract large predators to the coast. It's a fear that's already gushed along the northeastern seaboard of the United States, following the recent death of a boogie boarder who was bitten by a great white shark off the Cape Cod shore, the first fatal shark attack in Massachusetts since 1936. Many believe that the growing population of grey seals has triggered the increase in sharks to the area and there have been vociferous, knee-jerk calls for seal culls as a result.

While climate change might, in time, bring new shark species to UK waters, which could theoretically take advantage of grey seals' movements in and out of breeding colonies, orcas are the large predators that are currently most likely to be glimpsed. One study has shown that sightings peak in the months of June and July around Shetland, for example, coinciding with the common seal pupping season, with a smaller increase later in the year at the greys' breeding time. Though this suggests that orcas are freely availing themselves of plentiful pinniped prey, it's worth noting that there's never been a documented killing of a human by an orca in the wild.

The tide, storming in to my right, seems to be invigorating the seals in this people-free part of the colony midway between Winterton and Horsey. Bulls plunge through the surf and emerge bearded with spume. A cow does her best to herd her pup up the beach but it's having too much fun wriggling in the swash of the waves, nosing the curds of foam. I drop into a dip just below the dune-top path and nestle in to watch while wind-driven sand pitter-pats against my coat and adds a gritty consistency to the tea I pour from my flask.

Almost directly below me, a cow's lying on her side, accompanied by two whitecoats. One is already suckling; the other's yowling and trying to latch on to her top teat. Unusually, she doesn't seem to mind the presence of this second pup whose baggy skin and persistent cries imply it hasn't been fed by its own mother, of whom there's no sign, for quite some time. I'm more used to cows biting interlopers or batting them away but this one gives a swift scratch of the pup's head with the claws of her foreflipper, a familiar bonding gesture, and allows it to start feeding.

Though I've heard mention of these fostering, or allo-suckling, events, I've never before been lucky enough to see one. It's not uncommon for separations between mothers and offspring to occur in crowded breeding colonies when there's so much tactical manoeuvring and bellicose behaviour between adults and, even if a cow's successfully bonded with her pup and learnt to recognise its smell and call, reconnection isn't guaranteed. Researchers' observations, however, suggest that the incidence of allo-suckling varies greatly between pupping sites and, though it's more likely to take place on the periphery of the colony where the younger and less experienced cows may be situated, it remains an atypical occurrence.

While it might be tempting to consider whether a double suckling could indicate that the cow has herself birthed both pups, grey seal twins are an almost unprecedented phenomenon. In fact, the only recorded birth of wild grey seal twins in the world was confirmed as having taken place here at the Horsey colony in 2015. When a Friends of Horsey Seals warden came upon a cow and two newborns in the dunes close to one of the roped-off viewing areas, with no other adult female nearby, he had the presence of mind to collect some post-birth samples of blood-soaked sand. The cow started to feed both pups but, possibly due to human disturbance, ended up leaving them after ten days. They were rescued, nourished to a healthy weight and, ultimately, released back into the wild, though not before hair from both pups, as well as the post-birth blood, had been sent to the Institute of Marine Research in Norway. DNA analysis verified that the pups, one of each sex, were indeed twins, fathered by different bulls.

I resume my walk above a narrow strip of beach where the tide's forced all the cows and pups up against the seawall. While the bulls still seem energised by the big waves breaking over their backs, the drenched pups are visibly sea-weary. Those lucky mothers who happen to have birthed close to one of the raked gaps in the flood defences have coaxed their pups up the slope beyond the tide's reach. I'm reminded that a leaked 2008 Natural England report argued that the sea defences along this very stretch of coast were unsustainable 'beyond the next 20–50 years' and, in recognition of the fact that maintaining or upgrading them would cost millions, one of the suggested policy options was realignment of the shore – in other words, let it flood and lose hundreds of homes, several villages, medieval churches and nature reserves in the process. If this scheme were ever to be adopted, the entire Horsey pupping colony would also, of course, be forced to relocate.

As I draw closer to Crinkle Gap, visitor numbers start to increase again. Before I reach the viewing area, I keep a lookout for a cow-and-pup pairing that I spotted on my wanderings a couple of days ago. The cow had left the beach and scrambled up into the dunes to give birth, and though the pup was enthusiastically suckling and all seemed serene, their close-to-the-path position made me fear they'd be vulnerable to harassment.

I had not, however, reckoned on their being interfered with to such an extreme degree. When I get to the hollow that the cow selected, I find that twelve or so visitors, some crouched and some standing, some with phones and some with more sophisticated cameras, have homed in on the pup like paparazzi and there's no sign of its mother at all.

'Look at it sucking its thumb – it's adorable!'

'Such a cutie!'

'It may look cute,' I blurt out, 'but it's so hungry that it's trying to suckle its own flipper – it's desperate!'

I try to recall what Gemma said about the Friends of Horsey Seals policy on rescues. When a pup's abandoned in the so-called 'honeypot' area of the colony – the part which receives the most visitors and is visible from the two viewing platforms – it'll be monitored for a while to see if its mother returns and if she fails to do so, a choice will be made as to whether a rescue should be effected. 'If it's a newborn, we only have about twenty-four hours to make a decision. Three days if it's had some feeding and gained some blubber,' she said before hinting at one of the factors on which the decision may be based. 'It takes a mum three weeks to wean her pups – it takes a rescue centre at least three months and a lot of resources.'

The pup lets out a pitiful cry.

'Where's your mummy gone, darling? Sorry, sweetheart, I can't help you.'

'Of course you can help – we can all help!' I say. 'Leave it alone. Stop harassing it. Its mother's gone because she couldn't deal with all this disturbance but if we move away, there's still a chance she'll come back.'

When it becomes clear that, without a hi-vis waistcoat on, I've little hope of getting anyone to pay heed, I prise myself away from the site of the struggling pup and hurry on to Crinkle Gap to seek out a warden. I'm grateful that the first one I speak to summons the duty warden and, over the next couple of hours, I listen in on various walkie-talkie and in-person discussions of the pup's welfare. Friends of Horsey Seals are already aware of its solitary state and believe its stressed mother left yesterday. Now that there's explicit evidence of further, human-induced disturbance, the situation's deemed to be more critical. The duty warden consults with RSPCA East Winch, the wildlife centre about an hour away where pups from the Horsey colony are rehabilitated, and Peter, the long-standing chair of FoHS, arrives. Finally, having taken into account the fact that the abandoned pup is easily accessible and no other breeding seals will be disrupted if a rescue takes place, the decision's made to delay no longer and move in.

It's late afternoon and the sun has dipped low behind the dunes by now – the whole beach is in shadow and only the topmost tips of marram grass are drizzled in light. Several hundred pink-footed geese fly over, heading for their evening roost after a day of grazing on sugar beet remains. Their continual, in-flight yapping calls are considered to be convivial, uttered by family members keeping in constant contact, but today, they come across as nervy and urgent.

As the goose tumult wanes, Peter plus one of the wardens make their way over the dunes to the hollow that houses the hungry pup. I watch from a distance alongside a huddle of visitors that includes a child with pretend binoculars made from two taped-together toilet rolls. Having covered the pup's head with a towel to keep it calm and protect themselves from being bitten, they manoeuvre it onto a stretcher, fold up the sides and carry it down to the Seal Rescue vehicle that's waiting on the track behind the dunes.

'Our chairman used to use his own car for rescues,' I'm told, 'but his wife got fed up of it – they're quite pongy, seals – so we have a proper vehicle now.'

Feeling a mix of sadness and relief as the pup is driven away to East Winch, I head back towards Horsey Gap and clamber up to the viewing area for the final time. Beyond the dunes, the full moon is rising, mottled and huge like a pregnant cow. Having grown briefly intimate with both the fragilities and vigour of another remarkable east coast seal site, I'm now going to be turning my attention west again. Soon, I'll be seeing Dad and, while his heating timer issues seem to have been resolved, I'll need to help out with all the other domestic and admin tasks that have mounted up while I've been away. There are medical appointments looming too, including an echocardiogram for a routine check of his heart murmur.

In the deepening dusk, I watch a body-slapping bull, the thump,

thump, thump like the beach's reliably beating heart. Two silver-coated weaners cautiously approach each other, meeting nose-to-nose before swerving away.

And then, just as I'm on the very point of leaving, I finally see one. After all those days of waiting and near-missing, not just here but at Donna Nook and Blakeney too.

No restlessness.

No circling.

No digging.

One minute the cow's rolling onto her belly from her side.

And the next, a pup, its white coat stained with blood and yellow fluids, has burst from between her rear flippers through amniotic sac onto sand.

22

Hauling Out

'Dog.'
  'Cat.'
  'Mouse.'
  'Cheese.'
  'Cheddar.'
  'Gorge.'

A few days on from my return to Wales from Horsey, I'm sitting with Dad in the cardiology department of the hospital. We arrived at 9 a.m. for his echocardiogram: his heart was scanned at 9.30, after which he was invited to plonk himself down on a blue plastic seat in the busy waiting area. It's now past midday, I've already fetched us tea from the café, we've read a newspaper from the hospital shop and have moved on to playing a word association game to ease his frustration at the wasted time.

  'Gorge'
  'Scoff'
  'Mock'
  'Exam.'
  'Pass.'
  'Fail.'

Like juvenile seals with their tireless aquatic antics, Dad is happy to play this game for hours and, in spite of his memory challenges, is still able to make quick and creative connections between words.

'Fail.'

'Fluff.'

'Down.'

'Up.'

'High.'

'Society.'

'Mr Richardson? Sorry to have kept you waiting.'

When a doctor finally appears, Dad sighs with irritation at the interruption.

'My name's Dr Abbas. I'm part of the cardiology team and I need to let you know that there are a few issues with your echocardiogram.'

'What d'you mean?' I say. 'What sort of issues?'

'I'm afraid we're going to have to keep you in for a little while.'

'In where?' says Dad.

'In here, Mr Richardson. In the hospital.'

'What issues? What's wrong with him? I'm his daughter—'

Dr Abbas's eyes do a circuit of the waiting area. 'Can I ask you both to come with me, please.'

Dad tuts as he gets to his feet, still more concerned with returning to our game. Having followed the doctor into a side room that offers more privacy, I perch on the edge of another blue plastic seat, legs madly jiggling.

The echocardiogram couldn't reveal the status of Dad's heart murmur and how the aortic valve is performing, Dr Abbas at last imparts, as his heart is beating erratically and much too quickly. The condition, known as atrial fibrillation, can cause a number of symptoms including shortness of breath. 'We'll do some further tests and get you started on some medication right away. As soon as we stabilise your heart rate, we'll be able to do another echo and have a look at how the valve's doing.'

My own heart has started racing in response to his words, as has my flustered brain. I noticed Dad was short of breath when I phoned him from Horsey on the day he was confused about his boiler controls. If only I'd been more aware – I got bogged down, though, in worrying that he'd be cold without his heating on instead of realising something more serious was amiss. The one good thing is that I can at least support him through this hospital admission instead of having to make a cross-country dash back to Wales from some distant seal site as I had to do earlier in the year.

'Is this okay with you, Dad? Are you happy with what's been decided?'

He frowns. 'Nothing wrong with me, as far as I can tell. But I suppose they know what they're doing.'

'One of my colleagues has just confirmed there's a bed available,' says Dr Abbas. 'I'll get her to take you to the ward.'

'How long d'you think he might be in for?' I'm wondering how big a bag I'll need to pack.

He gives the first half-smile of our conversation. 'There are few certainties in medicine. But he could be home in four or five days if all goes well.'

Two weeks on, I'm taking the now-familiar route to Dad's ward for my daily visit, passing the listing tinsel Christmas tree at the entrance and greeting some of his fellow patients. The man whose bypass surgery has been cancelled four times. The man who refuses the hospital food and survives on his wife's supplies of soup, orange squash and pasties. The man who needs supplementary oxygen and always lends Dad a pen when he loses the one he's using for the crossword. The man who wheels the communal TV to the bottom of his bed and works himself into a red-faced lather at the rugby or football, a perilous pastime for a patient on a cardiology ward.

'Ready to go?' says Dad.

As has been the case most afternoons recently, he's waiting with his bag packed. He doesn't, however, speak with the feisty *I'm going home whatever anyone else says* insistence that he displayed back in the summer but rather from the genuine belief that, earlier in the day, some doctor, nurse or specialist discharged him.

'Sorry, Dad. Not today.' My words inspire a perplexed expression which intensifies as I unpack a few items from his holdall – slippers by the bed, washbag on the bedside cabinet, pyjamas on the pillow. As usual, I also track down his towel in the communal bathroom and return the one he's packed by mistake.

'What's happening now? What are you doing that for?'

'Just making sure you've got everything you need, Dad, that's all. Now, how about we play a game? I've brought snakes and ladders and dominoes today.'

His eyes light up as soon as I mention the latter. While he hasn't felt like reading any of the books, either his favourite nature writing or poetry, that I've brought along, playing a board or word game still seems to calm and uplift him.

With huffing effort, he pulls his wheeled table closer to his chair, moves aside the water jug and plastic glass and waits for me to tip the dotted tiles from their box.

'How many do we take? Seven, isn't it?'

I, too, have been finding that absorption in games offers brief but welcome respite from worrying about Dad's health. It took longer than expected to stabilise his heart rate and when the echocardiogram could eventually be attempted again, it showed that his heart murmur has worsened and the aortic valve narrowed to a 'serious' degree. However, since only an upgrade to 'critical' would qualify him for immediate surgery, he has to wait at least four months for the operation that could help him. In the meantime, he's growing increasingly breathless, as well as bloated due to fluid retention, another red flag symptom of heart issues. Last week, his feet, ankles, knees and thighs became swollen and now his belly's engorged too. I'm alarmed not just by the fact of his illness but also by the rapidity of its progression.

Having selected seven tiles, he gives lengthy consideration as to which to place on the table first. When the game gets underway, he has a speedier strategy, impatient to add to the dotted line and cock-a-hoop when I'm unable to add a tile of my own.

We're likely to have at least another week of game playing during visiting hours ahead of us. After umpteen conversations with the ward's rotation of medical staff, I've been told that a range of tests, including ultrasound scans and an angiogram, will be carried out to determine if he'll need open heart surgery or whether a less invasive procedure will be possible. Once all the information's been assembled, the cardiac team intends to discharge him so he can wait at home for an operation date to be confirmed. While returning home might ease his disorientation and confusion, living there with an accelerating heart condition and no medical back-up feels like an enormous risk.

'I win!' Dad places his last tile, a double blank, on the table with a big wheezy grin. 'Best of five?'

The man in the bed next door is a gentle soul who periodically lapses into French – *'Bon appetit! Merci mille fois!'* he shouts whenever a healthcare assistant comes to feed him – and is convinced that the nurse who administers his medication is the daughter of whom he's tearfully proud.

By contrast, Dad's prevailing mood is neither gentle nor grateful. The New Year brought not the expected discharge and extended wait at home for a date for his heart valve operation but, due to his worsening symptoms and need for continued care, a move to a different hospital. And here, rather than being based in a cardiology unit, he's been assigned a bed in a dementia ward.

Some days, he refuses to swallow his diuretic tablets because he believes the staff are trying to poison him.

He leaves his shoelaces undone because he can't reach past his distended belly to tie them and, in spite of the tripping risk, he won't let either me or a nurse assist him.

He's torn and shredded the pages from his crossword book and flung all the pieces across the ward.

He points out imaginary surveillance cameras, complains he's constantly being watched. At night, he rages against going to bed and wanders up and down for hours so he can stop the staff from harming him while he sleeps.

Once, pretending that he needed to use the bathroom, he hid in an adjacent storeroom and was found fifteen minutes later, curled foetally between piles of toilet rolls and paper towels.

Most days, he glares at me with hatred and suspicion. If I try to retrieve something from his holdall, he howls with the vehemence of a cow seal who's had her personal space invaded. He believes that I'm leading a conspiracy against him and that every doctor and nurse is doing my malicious bidding. I've learnt to compose my face into as neutral an expression as possible as any attempt at cheerfulness generates more anger. 'What are you laughing for? Think this is funny, do you?'

At the end of each visit, I crawl from the ward, shaking with shock and despair. How can he have become so ill so quickly? How, in the space of six weeks, can he have skipped several stages of dementia and hurtled into this aggressive, super-paranoid phase? The nurses display a we've-seen-it-all-before impassivity. The latest consultant to whom he's been assigned, a specialist in geriatric medicine, can't offer me an appointment for at least another week, and there's no up-to-date input from a cardiology team either. While letting out a torrent of sobbing would be a good release, I hold it all in, push on, push through, knowing that if I were to start to cry, I'd never manage to staunch it.

I try rattling the box of dominoes, hoping the sound might jolt him back to a semblance of who he used to be. I try reading aloud, just as, through my childhood, he read aloud to soothe me. I bone up on dementia resources and even find myself begging the ward manager to try to get hold of an innovative – if improbable-sounding – therapeutic tool. Introduced into a range of elderly care settings, PARO's a cuddly robot that happens to have been created in the form of a baby seal.

Through artificial intelligence technology, PARO learns to behave in ways that each patient favours and to respond to its name. By virtue of built-in sensors, mics and motors that reposition its flippers and head, it reacts to being stroked and spoken to with movements and pup-like vocalisations. The decision to make the robot a seal apparently stemmed from the fact that people rarely have unhelpful associations with real pinnipeds while its exaggeratedly cute baby creature appeal seems to empower patients to want to give care as well as receive it. Hard though it is to believe that Dad's demeanour would radically change on interaction with a robot, PARO's consistently been found to lessen dementia patients' agitation, so I'd give anything for him to have the opportunity to try it out. I'm made aware, however, that while it's been widely employed in Japan, its place of origin, and in the US and parts of Europe for well over a decade, there hasn't been extensive uptake in the UK, perhaps because of its £5,000 price tag. Nevertheless, my fruitless pleading for PARO has a positive outcome. Having babbled on about the benefits it might offer, I win permission for Dad to therapeutically connect with a way more engaging live animal – Hooper's welcomed to the ward instead.

Hooper's not banned from rummaging in Dad's holdall as I am – he's allowed to snout around in it and steal his socks. He locates a slipper under the bed and shows Dad what he's found with a wag that's not confined to his tail but animates his whole body. And when he settles close to Dad, head on one of his swollen feet, Dad's face loses its scowling tension and remembers how to smile.

My own smile doesn't even hint at a return until we finally meet his new cardiologist, a heart valve replacement specialist. He's personable and reassuring, insisting that Dad's drastic behaviour change doesn't surprise him. He's witnessed this in cardiology patients before, with the heart's failure to pump blood to the brain efficiently being a contributing factor, and doesn't believe it's an irreversible symptom of rapidly-progressing dementia. He also has another look at the first hospital's assessment of the status of the valve and immediately reclassifies it as 'critical', with the aim of intervening as soon as possible. The burden of fluid that Dad's carrying, though, has to be radically reduced before any operation can take place: this requires the diuretic tablets to be ditched and a much higher, round-the-clock dose to be administered intravenously.

At first, Dad keeps ripping out the tube from the vein in the back of his hand and a date for the valve replacement seems doomed to remain as distant as ever. Thanks, however, to one extra-persistent, patient nurse, who somehow manages to develop a knack for pacifying him, he gradually becomes more trusting and compliant. Once the treatment's properly underway at last, his daily weigh-ins show that the fluid starts to decrease quite quickly like the blubber of a mother seal in her weeks of intensive pup-feeding.

And, while my anxiety doesn't ebb away at anything like the same rate, I at least dare to believe I can begin to breathe a little more freely.

It's a grey, late January day and I'm finally managing to make for the coast path at Aber Felin for the first time since early November. The fields through which I'm squelching are cratered with quad bike tracks and sheep's hoof prints, all retaining the past days' rain. I sidestep occasional splashes of yellow fungus too, which I recognise as the aptly named dog vomit slime mould.

When I reach the coast, a robin monitors my stuttering progress. It feels odd to be out on the cliffs again – I know this track intimately but my feet are faltering, currently more attuned to hospital wards and corridors than rocks and tufts of grass. The past week of hospital visiting's been a little easier in that Dad's cognition and mood have both continued to improve as more of the fluid has been offloaded but I'm still churning with concern as we wait for the valve replacement date. When Dad's favourite nurse advised me to have a break from visiting and get some rest for a few days, I dithered and quibbled but finally acquiesced to an overnight trip home to Pembrokeshire, persuaded by the prospect of a check-in with the Aber Felin seals' winter routine.

A few months on from pupping and mating, grey seals shed and renew their fur, spending the bulk of their time ashore during these weeks in crabby congregations, sometimes at the birthing sites and sometimes at dedicated moulting locations. I've always found it fascinating that some pupping places transition into winter haul-outs while others don't, and that some are preferred by adult males while others are favoured by females. As far as Aber Felin's concerned, although the bay's first two beaches are popular pupping spots, seals rarely frequent them at any other time of year, while the beach below the ledge lookout evolves into a cow-dominated moulting and haul-out site between November and March. Various theories have been put forward to explain what triggers the moult's annual onset, including endocrine processes such as the hormone levels associated with cows' reproductive cycles, and photoperiod – seasonal changes in day length and the hours of light the seals are receiving. Other factors may influence the moult's duration, including ambient air temperature and the fitness of the individual, largely based on the past months' feeding success. Overall, however, in contrast to the plenitude of research that's been undertaken into grey seals' breeding behaviour, there have been relatively few studies of their moulting activity.

As I edge my way onto the ledge above the third beach, I feel uncharacteristically vulnerable. I even feel awkward using my binoculars: they don't seem like an extension of my own eyes any more. Gradually, though, I start to tune in and manage to count over a hundred cows harrumphing and bickering their way through their moult. Their fur looks as patchy as a threadbare carpet or a rug with cigarette burns. A number have neck and back wounds from over-zealous scratching – there's an increase in the flow of blood to the skin's surface as well as to the hair follicles during the moult which makes the seals more sensitive to the cold of the sea and also more prone to bleeding. Hauling out on the beach reduces heat loss but rubbing against rocks to relieve itching inevitably causes minor wounds.

In addition to all the scratching, there's some gargantuan yawning and more delicate uncurling and fanning of rear flippers. Each time one cow moves, it triggers a ripple of howls and bared teeth and the need for every seal to reclaim some personal space. They're even tetchy with the tide, grumbling as it inches in, time and again collectively shifting further up the beach. A couple of high-spirited juveniles, who are also visibly in moult but must have failed to get the memo about the need to conserve energy and heat, are vexing the resting cows too. They surf onto the beach on an incoming wave, wrestle on the pebbles, then plunge back into the sea when a cow lashes out, only to allow themselves to be washed up again.

Before I head home, I decide to walk on to the low, narrow promontory where I picnicked with Hooper on that golden July evening last year. I pass a clump of bluebell bulbs, unearthed by the raising of a molehill, and hear a great tit's squeaky bicycle pump song, two intimations of a spring that I can't find it in me to believe in yet.

Leaving the main path, I pick my way along to the headland's slippery tip and try to take some deeper, more easeful breaths there.

A speckled-breasted rock pipit hops from crag to crag.

Shrill trill of an oystercatcher.

A weave of seaweed unravels on the surface of the water, disturbed by the emergence of a seal. She's a juvenile and she's spotted me, lifts herself a little further out of the water, craning her neck for a better view. The sea's still enough for me to see her wafting her front flippers from side to side as she maintains her upright position. She's scenting me too – the parallel slits of her nostrils are flaring but though I likewise sniff the air, my far less sophisticated sense of smell relays no information. I have to rely on my sense of sight instead, taking in her prolific whiskers and matching eyebrows, and reading the dark grey blotches of her neck fur like a Rorschach test. Having met each other's gaze, we seem to be seeking to maintain it and I feel I'm being thoroughly seen for the first time in weeks.

As dusk starts to steal in, I let myself continue to be seen and scented, and wonder which of us will be the first to look away.

Even just a few hours of communing with the seals boosts and reboots me and enables me to face the valve replacement phase of Dad's illness with fresh energy and perspective. Having had confirmation that he won't be subjected to open heart surgery, I learn more about the alternative procedure whereby a catheter's inserted into an artery in the groin under local anaesthetic and the new valve is remarkably sent heartwards. While it's less invasive than surgery, it of course still carries significant risks.

It also requires Dad to be transferred to hospital number three and this time, he's only mildly disorientated by his new surroundings. What's more, he's pretty unfazed by the procedure itself: while I spend the day immersed in a whirl of distraction activities and am drooping with fatigue by the time I visit him in intensive care in the evening, he's sitting up in bed, ready to chat and savouring his supper of macaroni cheese and ice cream. 'They wouldn't let me eat all day,' he says. 'And I've built up quite an appetite.'

Over the next few days, he continues to make gentle progress, moving out of intensive care and onto a regular cardiology ward. He's less breathless and, mood-wise, perkier than he's been in a long while. I no longer approach visiting times with the same sense of dread and, in the hours between visits, feel more able to give my attention to something other than Dad's physical and mental health. Hooper's walks get longer, I start to tackle my backlog of mentoring work, draft a poem in preparation for this year's British Animal Studies Network residency and plan a trip home to Pembrokeshire that lasts for longer than one night.

When, a week on from the operation, my phone rings mid-morning, I'm expecting it to be one of the mentees to whom I've promised some verbal feedback on a piece of writing.

'Is that Susan?'

'Yes.'

'This is Abigail, the ward sister from cardiology.'

Instant gush of adrenaline.

'I'm afraid your father's had a collapse.'

Legs wobble and voice follows. 'S-sorry?'

'Are you able to come into the hospital?'

'I don't . . .' Swallow. Try again. 'I phoned first thing . . . he was fine . . .'

'He had to be resuscitated, Susan. We need you to come in right away if you can.'

All the clichés – urge to retch, galloping heart, clammy palms. Flashback to the shock of Mum's sudden hospital demise. 'Is he . . .'

'I'm not able to give out any details over the phone, I'm afraid. We'll tell you more when you come in.'

Twenty minutes to get to the hospital. *Will I get there in time? Is he conscious? Will I have to say goodbye?* Then, endless corridors, a lift that refuses to respond to my frantic jabbing of the buttons, flight after flight of stairs. A nurse in green intercepts me as I burst through the ward's double doors, guides me into a side room. A few chairs. A table. An ominous box of tissues though I feel too shocked to cry.

'Hello, Susan.' Another nurse, navy blue with an Abigail name badge, and a doctor, tame beard, wild brows. 'Can we get you something to drink? Tea? Coffee?'

I shake my head.

'Water?'

'Just tell me about Dad. *Please.*'

They sit opposite me, she leaning forwards, his eyes down on his notes. He explains that earlier this morning, Dad got up from his chair and the man in the bed next door saw him crumple forwards onto the floor. Medical staff raced to his aid, couldn't find a pulse, had to restart his heart. They've since been working to try to stabilise him and establish what caused the collapse. The valve replacement consultant's been summoned and will be seeing Dad later, a cardiologist who specialises in the electrical system of the heart is involved now too and lots of tests are being run.

'Can I see him?'

'We'll take you there now.'

He's back in intensive care, wired to a heart monitor and an oxygen supply. Understandably, he's sleeping, though when I take his hand and stroke it, his eyes flicker open and a smile glimmers beneath the clear plastic of the oxygen mask. I'm allowed to stay at his bedside, talking, as he mumbles in and out of sleep, about Hooper and birds and crosswords, twitching each time the heart monitor beeps and willing there to be no decline whenever a nurse arrives to check his vital signs.

Rescued. He's been rescued. As the day lurches on, my mind reels through all the arguments I've heard over the course of my seal journey about when and whether to rescue a suffering pup. Only rescue if it's a human-induced trauma or if the pup's in public view. Don't intervene at all – leave it to nature. Rescue only if there's available sanctuary space and funding. Spare the pup undue stress and euthanise instead. And the strategy that most closely matches Dad's alacritous rescue – if the pup's in distress and accessible, step in right away.

At last, the electrocardiologist appears and, after studying Dad's medical history and today's test results, he recommends that a pacemaker be fitted. Dad's fully emerged from sleep by this time and, having no memory of what happened earlier, is baffled by what he's hearing as well as by the sprained wrist that he sustained from his fall.

He's baffled by my continued presence too. 'Are you still here?' he says when, propped up in bed and brighter than I, or anyone else, expected, he starts making inroads into ocean pie and mash.

'Six hours and counting, Dad,' I tell him.

Hauling Out

Now that the oxygen mask's no longer needed, I get the whole panorama of his smile. 'Soon as I've finished eating, d'you fancy playing a game of something? How about the one where we connect all the words?'

Mid-March, approaching the spring equinox, and it's the warmest day of the year so far with barely a riffle of a breeze. I walk my shadow along the coast path at Aber Felin, watch it fall across celandines, a sunning red admiral, a drift of blossoming blackthorn. In the field to my right, leggy lambs are suckling, their yet-to-be-docked tails still free to wag with the glee of it, while the sea to my left seems almost entirely static, the waves barely lapping, let alone building themselves up to break.

My brain has yet to attain equivalent clarity and calm in spite of the fact that, after all the vicissitudes of the winter, Dad's health at last seems to be on an upward curve. It's almost three weeks since his collapse, two since he had the pacemaker fitted and the specialist's pleased with its regulation of the heart. While Dad's still experiencing some fluid build-up, it's mostly confined to his ankles and calves and is nowhere near as extreme as in the months before his valve replacement. The cardiology team is still trying to find the right diuretic medication balance – too little and the fluid build-up will increase; too much could damage his kidneys – but overall, they've been speaking positively about his condition. At the end of my last few visits, he's been able to shuffle down the ward with me, with the aid of a stick, as far as the exit. He's also started to read books again and is increasingly eager to talk about them. In the next few days, I've been told that an appointment with a discharge nurse will be arranged to discuss his post-hospital options, a conversation that, even as recently as last week, I never dared believe would take place.

The first two beaches are still unfrequented by seals and, though I have yet to access the ledge for the best view of the third, I can see that it's passed its winter peak. There are only twenty-two huffing and moaning cows hauled out, plus a rare sleeping bull, double chin flat on the pebbles. As was the case on my last visit around seven weeks ago, several juveniles are also bombing in and out of the shallows.

Having spotted a few dropped tissues and Werther's Original wrappers, I know that coast path visitor season is getting underway and a girl who looks to be about eleven years old joins me now at the cliff edge, beaming bracily. 'Are you here to count the seals?' she asks, craning to look in my notebook.

'Clover, come away from the edge,' says her father before deciding he needs to give a running commentary on the seals' behaviour ('That one's on the move, look!') as well as count them on my behalf ('You've got one, two, three, four . . . oh, at least eight down there).

'Why are they making that weird noise?' asks Clover, in response to the cows' moaning. 'Is it some kind of existential howl?'

I stutter in my note-making, so taken aback am I to hear an eleven-year-old reference the existential.

'Last year when we were in Namibia, we saw thousands of seals,' says her father. 'And oh, the smell.'

Today, the prevailing smell isn't seal but the coconut-and-vanilla blend of sun-warmed gorse. A distant chiffchaff repeats its name over and over, only to be drowned out by the drone of a little fishing boat entering the bay. A man in orange oilskins starts lowering lobster pots over the side – he works quickly and methodically with not even a glance in the direction of the seals and though they register his presence, he's far enough away to cause them no concern.

While my binoculars have been focused on the fishing boat, a young guy in a safari-beige gilet has replaced Clover and her father, who've resumed their enviable daughter–dad walk along the coast. He delves into one of its multiple pockets for tobacco and Rizlas and starts rolling a cigarette. We watch a juvenile hurl herself onto the beach, setting off a domino effect of grumbling and flipper-flapping among the adult cows. I point out the bull, the only seal who hasn't bothered to rouse himself from sleep, and a look of shock flashes across the guy's face. 'I don't mean to be rude but are those all his women?'

I eventually make my way off the main path and onto the ledge viewpoint from which I'm able to see a broader area of the beach, including the very back of it close to the base of the cliffs. A mesh of marine debris looks to have accumulated there – though much of it's tucked behind a rock, I clock some thick blue rope, a red Croc, an obscure pile of off-white cloth. When I check through my binoculars, however, I'm startled to find that it's not cloth at all but the rear flippers and lower body of a half-hidden pup.

Out-of-season births are not unknown but this is the first time I'm aware of one occurring at Aber Felin. I wonder if the delayed implantation that's characteristic of grey seal pregnancies didn't happen in this case for some reason. Following a successful mating in the autumn breeding colony, the cow's fertilised egg divides, at which point pregnancy is suspended for three and a half months. This allows the cow to feed and regain weight and condition, after which there's an eight-month gestation period, resulting in her giving birth at the breeding colony one year on from the previous season. If implantation fails to be delayed, however, a spring birth will occur.

I scan the beach more carefully to see if I can match a mother to the pup but there's no obvious candidate – no close-by cow who's acting protectively, no sign of lactation as far as any of the adults who are lying on their sides are concerned. The pup isn't behaving as if it's been abandoned though – it isn't wailing with hunger or dragging itself desperately across the pebbles and it's definitely alive as its rear flippers are now flexing and stretching.

I'll be leaving Pembrokeshire again later today to resume my hospital-visiting routine so I'll have no way of knowing if this unseasonal pup manages to survive. I feel unexpectedly rocked by this life-or-death uncertainty, though in the context of the past few months, I guess I shouldn't be so surprised.

As a scream of oystercatchers streams by, slumped on the brink of this giddying cliff, I finally start to cry.

23

Moult

Since I was last here in Norfolk four months ago, the sand has been resculpted, stripped away in some places and built up in others by winter storms. A cosmopolitan scattering of marine debris has been deposited on the beach too – a milk carton, *1.5 % fett*, from Germany, a Maison Verte detergent bottle, *écologique*, from France – as well as the usual twists of rope and fragments of polystyrene. There's also been a shift in the bird life: while, in December, Horsey's shore was presided over by gangs of great black-backed and herring gulls, now that it's April, skylarks are most conspicuous, rising high from the dunes that line the back of the beach, bubbling over with song.

Even more striking than the bird and sand rearrangements is the change in the behaviour of the grey seals. No longer a breeding beach with hungry pups, mothers giving suck and libidinous males, Horsey's transformed into a haul-out for adult moulters. In contrast to the hundred or so cows that rested and kvetched on Aber Felin's pebbles during the Pembrokeshire moulting season, several thousand seals in mixed-sex groups are stretched out on the sand to Crinkle Gap and beyond, interspersed between the rock and timber groynes. Though cows enter moult in advance of bulls, there's a noticeable overlap between the two sexes' moulting periods here. The usual contrast in coat colour and patterns – the females pale grey with charcoal blotches, the males mostly darker – is less obvious as the dead fur they're in the process of shedding has turned many of them patchily brown. Overall, at this early hour of the morning, they seem quieter and less fractious than the Aber Felin cows, their bodies prone but for an occasional flipper lifted for a stretch, a neck raised for a scratch.

A number of youngsters, December's weaners, are lingering on the fringes of the groups of adult moulters. I spot one barrelling along the sand but when he tries to sniff another's nose in greeting, she jerks her head away like a child shying from a great-aunt's fusty kiss. Others are trying out adult behaviour, defending their personal space by flipper-flapping and making hoarse, unconvincing attempts at a howl.

After all that's happened over the winter, the fact that I've managed to get my agitated self back across the country to Norfolk feels nothing short of miraculous. Dad's continuing to make peaceful progress in hospital and, provided there are no setbacks, in a week or so when he's discharged, he'll be moving into a care home for a period of convalescence. In the months since I was here for the pupping season, I've been changed and rearranged just like the sand and, before helping Dad settle into his new environment, I'm hoping that spending a few days at the first large-scale moult I've ever seen could enable me to start to rest and reset.

As the sun hoists itself higher in the sky, restlessness ripples through part of the haul-out. A cow's head pops up like a meerkat's over the back of a dozing bull and after a quick look round, she ducks down again. Other cows follow suit, periscoping their necks, their anxious glances skittering between the beach and each other, an external manifestation of my own inner jittery vigilance.

The reason for this increased wariness is clear: at moulting time, there are no ropes confining visitors to the viewing area along the top of the dunes. Instead, while there are notices advising that we should refrain from getting too close, humans are allowed to access the beach here at Horsey Gap and share the sand with the seals. Though wardens are patrolling, there seem to be far fewer than when pupping was underway and, with it being the start of the Easter school holidays, the potential for disturbance looks to be high.

People are starting to use the big mound of sand against which I've tucked myself as a means of clambering down onto the beach from the path along the seawall. A man's tugging a toddler with one hand and, with infinitely more care, carrying the first ice cream of the day with the other. His wife's coaxing down a poodle in a pink onesie and Granddad is puffing at the rear. On reaching the beach, he brings his hands to the small of his back and frowns. 'Don't they put any deckchairs out?'

They join the nearby semi-circle of seal watchers, all of whom are so far keeping a solicitous distance. A child announces that he's named a few of the seals – 'The one on his back is Roly. And the one that keeps scratching is Fidget' – while a man is giving a bull a human comedy voice à la Johnny Morris. Slightly apart from this main group, a woman narrates what she's seeing to her whippet – 'Look, George – there's a big old seal having a big old yawn.'

Strangely, a minority of visitors don't seem to register the seals at all – it's as if they've become as mundane and unremarkable as the backwash of a wave. Though there's a huge stretch of beach to choose from, I watch a family make a beeline for a spot no more than five metres from the edge of the haul-out. On the way, Mum stops even nearer to a winter weaner to tip sand out of a whingeing kid's wellie. Having parked their baby on a blanket, Dad starts fighting with a windbreak while the oldest child, a girl with a waist-length plait, clambers up onto the rocks of the adjacent groyne, inspiring her grandmother to shout, 'Fiona – you're an EXPLORER!'

Their presence sets off the first of four stampedes that I witness over the course of the next hour. The second is caused by a man who, egged on by his mates, starts to honk and clap like a performing sea lion. Stampedes three and four stem from folk creeping too close to the seals as they seek to capture ever more intimate photos. Most of these trespassers shriek and go into shocked retreat as soon as the seals make their mass escape into the water. One woman tuts with irritation at the fact that they're 'so camera shy'. Another merely gets her child to wave bye-bye.

I also become aware that a small but vocal number of people seem less than delighted by the existence of the haul-out, including the woman with the poodle, whose pink onesie is now coated with the seal shit in which he's just gleefully rolled.

'So unhygienic . . .'

'Think of the diseases . . .'

'. . . fouling all our beaches . . .'

All uttered while standing next to the tremble of plastic bags and empty crisp packets that have accumulated on the windward side of one of the groynes.

Hoping that, being further from the car park, the seals beyond Crinkle Gap are less prone to harassment, as was mostly the case in the pupping season, I troop along the sand, hugging the seawall. In spite of the impression gleaned by one of today's visitors ('Is that all they do? Lie there and get rid of their winter fluff?'), the energy expended in shedding and re-growing a full coat of fur is high, and repeated anthropogenic disturbance adds a particularly adverse burden during this moulting phase. As I've already learnt, a seal, when triggered to be vigilant, quickly starts to mobilise energy in preparation for fight-or-flight. This means that physiological processes which are not required for immediate survival, including those that drive the moult, are suppressed and energy stores that may otherwise be needed for optimum breeding and feeding success are drawn upon. During the moult, high skin temperature, with blood circulating close to the skin's surface, needs to be maintained to maximise fur growth too, and recurrent stampeding into the water results in a rapid loss of body heat. At this point in the seals' annual cycle, therefore, persistent disturbance has both a short-term and a more protracted impact – hampering of hair growth and prolonging of the moulting period, plus chronic energy depletion with implications for the seals' long-term health as well as that of their subsequent pups.

Using the rocks of a groyne as cover, I settle down to watch a fifty-strong group of bulls and cows towards the southern end of the haul-out. The males, who are higher up the beach, are motionless as boulders and while the females at the edge of the sea are more unsettled, it's due to the incoming tide rather than because they've been spooked by humans. In contrast to the bulls, who seem to be at an early stage of their moult, with just some fur thinning on their noses and round their whiskers and eyes, the cows' coats look thoroughly mangy but they're mostly free of the wounds that afflict the Pembrokeshire seals, sand being so much softer than stones when trying to shed hair and quell itching.

A few small flocks of birds are pattering around the seals here too, another reflection of the reduced presence of humans. Turnstones, their black, brown and white plumage scruffily breeze-blown, rootling through a little patch of pebbles. Sanderlings in one of two modes – either scuttling at full throttle or standing stock-still on a single leg. Ringed plovers with their black eye-mask markings like a band of winged bandits.

With binoculars, I look deeper into the haul-out, as well as along its far margin where a few four- and five-month-old young-sters are scooting, hindering their elders' rest and getting shooed away. One, though, is much less mobile than the others and when I focus more closely, I see that she's trapped in the tangle of a white plastic sack. Too heavy duty for a domestic rubbish bag, it looks like an empty version of one of the polypropylene sandbags that get piled up outside vulnerable coastal properties to reduce the impact of flooding. As is so often the case with young seals, curiosity must have prompted her to investigate, and swim into, this unfamiliar object with its frayed, tide-torn openings. From this distance and because the sack's shrouding much of her body, I can't tell if she's underweight on account of it impeding her ability to forage and feed. While her left front flipper's unfettered, the right looks to be quite constricted.

I decide to report her predicament to one of the Friends of

Moult

Horsey Seals wardens and hurry-stumble through the soft sand of the upper beach back towards Horsey Gap. The duty warden to whom I speak assures me that they're already aware of her and are monitoring her as closely as possible. She's still fairly robust, able to enter and exit the sea at will and swim quite considerable distances considering her encumbrance. The hope is that, as long as she's isolated from the main haul-out and there's no risk of a mass flushing into the sea, a rescue team will be able to detain her and cut her free. However, though she was spotted alone several miles up the coast a few days ago, by the time rescuers had reached the beach, she'd regrettably left again.

Over the remaining couple of days of my Horsey visit, I keep my eyes peeled for the sandbag seal but fail to see her again. In spite of the vigorous wind that starts to gust in off the North Sea, I spend up to ten glorious hours a day at the haul-out, mostly at the more peaceful Crinkle Gap end. I delight in tuning in to the moulters' diurnal rhythms, observing how their behaviour changes in response to tide and weather. From low-water lethargy to the shift and shuffle occasioned by incoming waves and the lengthy episodes of play that seem to accompany each peak of the tide, the moult is full of micro-dramas. There's mock-wrestling and fake posturing between bulls, a half-hearted, almost affectionate dress rehearsal for the more purposeful fights of the breeding season. Bulls and young cows engage in sparky encounters too: sometimes, a flirty female initiates the play, roly-polying provocatively on the sand or heaving herself on top of a bull's recumbent form. On other occasions, the bull sets the game in motion, seemingly pinning a cow down but allowing her to wriggle free as he never would at mating time. When she hurtles away, he flumps along in pursuit, both of them ploughing through the haul-out, provoking a Mexican wave of howling.

In addition to making the seals friskier, the wind, and the sand it stirs up, seem to make them itchier. Grains collect in the folds of the bulls' necks, cling to the wet round the rims of their eyes. One bull is mad with scratching, flinging himself from side to side on his back, thrashing and spasming. A cow arches into a yogic back bend, reaching with her foreflippers to scratch those at the rear.

Wind-blown sand grits my cheeks too. Each time I pour a drink from my flask, my tea is silted with it. Common whelk eggcases like scraps of bubble wrap are sent skimming towards the dunes. Spume shudders like jelly. Drifts of sand obliterate my footprints in minutes.

Though it drifts more gently, the joy I find in being here layers itself like sand over the stresses of the winter. Spending time in close proximity to several thousand seals who are collectively purging themselves of a year's worth of fur is as extraordinary an experience as counting pups at Blakeney and it's beginning to rebalance me even more successfully than I hoped.

Beyond the end of the haul-out, towards Winterton, the beach is a mix of sand and occasional patches of fine shingle. I riffle through the little pebbles, picking out those that replicate the colours of the seals' pelage. In the months ahead, when I'm back in Wales and no doubt torpedoed by Dad-related anxiety again, I hope that cupping these pebbles in my palm will help me return to the calm I've found here. I pocket a plain brown stone the shade of shedding fur. A bull-black pebble. A smooth grey ovoid with darker flecks, matching the pelt patterns of the cows.

And an off-white diamond for the young seal in the sandbag, struggling to moult her plastic coat.

Before I return to Wales, I have one more location to visit in Norfolk and, since there are no major moulting or breeding sites between here and the South West, it'll be the last stop I make on my round-Britain seal journey. There's a pleasing feeling of symmetry about it as, having started my travels at the Cornish Seal Sanctuary, I'm now ending them in sterile white overalls and over-sized wellies at another wildlife rescue centre, RSPCA East Winch near King's Lynn. Though I've had more than enough of visiting hospitals for humans, I'm glad to be getting a guided tour of this seal equivalent, which originated as an emergency assessment unit in 1988 in response to the phocine distemper epidemic. It's since become renowned for its care of sick and abandoned grey and common seal pups and, most notably, adults who've suffered life-threatening marine debris injuries. Having been so troubled by the plight of the entangled youngster, I'm hoping to hear about the specialised care the centre offers, as well as some of its successful rehabilitations.

Alongside manager Alison Charles, whose quietly authoritative manner reminds me of the nurse who won Dad's trust when he was at his most combative, I peep into a few of the cubicles lining the centre's isolation unit where all the animals requiring intensive care can be found. As well as an otter cub retrieved from an industrial estate far from any water course, I glimpse a poorly young seal suffering from lungworm, his breathing laboured and wheezy, which likewise catapults me back to the most distressing weeks of Dad's illness. Alison explains that in addition to treating him with drugs, she lets steam into his cubicle twice a day to open his airways and encourage him to breathe more easily, using an innovative mobile sauna device she fashioned from a wallpaper stripper. The temptation to croon soothing words in the sort of tone that I use on Hooper – 'It's okay, little one, you're going to be fine. Aren't you a beautiful boy!' – is enormous but she swiftly advises against it. 'We try not to talk to them or make eye contact. We need to keep them wild.'

As we disinfect our wellies after our brief visit to the isolation facilities, Alison outlines her contrasting experiences of rehabbing grey and common seals. 'They're such different animals to nurse. Commons are so much more placid – greys will try to dislocate your knee. And while commons that are sick may just give up, greys keep on fighting till their last breath.'

We continue our tour in an adjacent seventeenth-century barn, a vast space with a high, beamed ceiling. 'Everyone said I should make this the Visitor Centre but it's perfect for the seals. We don't need to heat it – it's warm in winter with the thick walls, and cool in summer. It's lovely for them.'

I can't think of a better use for a four-hundred-year-old listed building either. The barn contains a row of fenced-off shallow pools in which pairs of pups are continuing their recovery. Compared with the limp, sickly seal I've just seen in intensive care, they're alert and much more active. 'These guys are all self-feeding. Up to eight hundred grams of fish three times a day,' Alison tells me. There's the hint of a quiver of emotion in her voice when she adds, 'Transferring them out here and seeing them look so much better – it's what it's all about.'

When she indicates that two of the pups in the barn were rescued from Pembrokeshire, one from near Tenby and the other from much closer to Aber Felin, I feel unexpectedly moved myself. Because neither Bel at RSPCA West Hatch in Taunton nor Mallydams Wood near Hastings had space for any more rescued seals, they were driven here, all the way across the country, a journey of over three hundred miles. Having again bitten back the urge to use cooing baby animals talk, I suggest that this must have been a stressful experience for them, but Alison shakes her head.

'I don't have any qualms about pups travelling at all – they travel beautifully. You'll be driving along and there'll be snoring coming from the back. The staff don't travel anywhere near as well as the pups!'

Before their release back into the wild, all the rescued seals are nurtured through a final stage of recovery, transitioning into one of the larger, deeper pools in an outside courtyard. 'We feed them at random times so they don't know when it's coming and we feed them random amounts at this stage too,' Alison says, emphasising the importance of replicating, as closely as possible, the unpredictability of food supplies that they'll face at sea.

Though I'd love my abiding memory of East Winch to be the first outside seal I see, a soon-to-be-released young female, swimming and tumbling in the recirculation current in her pool, my visit takes a far more sombre turn as soon as Alison broaches the marine debris issue. She shows me to another outdoor pool occupied by an adult seal with a deep wound, a startling pink gash all along the back of her neck, so conspicuous above the water's surface as she swims length after length after length. When she was admitted to East Winch, having been rescued on Horsey beach, she was 'necklaced' – entangled in plastic monofilament fishing line. Almost invisible in the sea due to its low optical density, monofilament line is so strong and resistant to degrading by sunlight and saltwater that it could endure in the marine environment for up to six hundred years. 'I couldn't believe her neck when I saw it,' says Alison, recounting how the line had sliced through the seal's skin and into her flesh. 'The wound was so badly infected – the smell was terrible – and she was so debilitated.'

So far, her treatment has taken the form of antibiotics, painkillers, steroids, plus several sacks of salt in her water each day to keep the wound clean, over a four-month period. She's eating well and now looks to be a healthy weight. 'She's quite a character – understandably very nervous of us which comes out as pure aggression,' says Alison, a description that could equally describe Dad's hospital demeanour in the month before his heart valve was replaced. 'But she's improved immensely being outdoors where it's a bit more natural. That's what I mean about the resilience of grey seals – no other species would have survived such an awful injury.'

As I know from having spotted the sandbag seal, fishing gear isn't the only form of man-made pollution that's threatening the animals' welfare. In the course of my travels, I've also heard about seals being unable to free themselves from bikini bottoms, six-pack rings, polypropylene packing straps – with each differently sized loop representing a hazard to a different age group – and even about one hapless bull who got his head wedged in a bucket. However, out of the fifty-plus necklaced seals who've received treatment at East Winch in recent years, a significant number have been gravely wounded by a particular kind of frisbee.

The hole in a flying ring provides an inquisitive young seal with a new opportunity to play or probe for prey. When it first gets lodged around the neck, it may not pose much of a problem but, over the months, as the seal grows, the rigid plastic tightens and starts to cut ever more deeply into flesh. Eventually, the ability to feed is also compromised as the neck that lunges forward to seize prey can no longer be properly extended. Swallowing is restricted too and whenever a seal trapped in a flying ring is rescued, he or she is always found to be in a very malnourished state. Every flying ring-related incident to which Alison's had to respond has, she confesses, left her horrified. 'I've seen a frisbee so deeply embedded in a seal's neck that when it was removed, I actually thought her head would fall off.'

Of all the anthropogenic threats to which I've been introduced on my journey, these revelations about the scourge of marine debris have offered the most graphic insight into seals' suffering. It's something of a relief to turn my attention back to the seal in the outdoor pool whose fishing line wound is healing and who's swimming lustily through what Alison hopes could be the final month of her rehabilitation. East Winch has an excellent success rate as far as nursing entangled seals back to health, and releasing them back into the wild, is concerned, though Alison's aware that only a minuscule proportion of animals who are afflicted by marine debris are admitted here. Countless seals and other marine mammals must be suffering, undetected, and dying, at sea.

It's not always possible even to rescue all those entangled animals who've been identified, and are being monitored, in and around the Horsey haul-out. In addition to the young sandbag seal, both the RSPCA and Friends of Horsey Seals are aware of at least eight adults who are caught in lost or abandoned fishing gear, and sometimes it can take several months for a suitable rescue opportunity to arise. Capturing an adult, even one who's apparently debilitated, is a much more challenging prospect than rescuing a pup and, on leaving the outdoor pool area, we return to the main building so that Alison can show me a key piece of equipment. Seals can't be darted as the effect isn't instant – they'll escape into the water and, when the drug takes hold, drown. 'So we use this instead,' she says, brandishing a long, tapered net attached to a large hoop. 'I bugged the Sea Mammal Research Unit for years to let me go out and practise with them and eventually they said I could. You have to run as fast as you can towards the seal you want to catch – someone from each side – and as it's moving off, you throw the net over its head. Once you've got it in there, you pull the drawstring tight around its hips and at that point you feel you've got some control – it stops moving. The day was great fun but scary – you've got to have a lot of nerve.' Her face breaks into a smile. 'And I just love that you can catch an adult seal with a hula hoop! It's mad!'

This is not the image with which I was expecting to end my visit to East Winch or, indeed, my entire round-Britain seal journey, yet it nevertheless feels appropriate. From Sue Sayer in Cornwall to Alison here in Norfolk, I've encountered grey seal advocates who are working with exceptional levels of spirit and ingenuity to mitigate the many challenges that the animals are facing.

From rehabbing seals, I now need to rechannel my focus onto my soon-to-be-care-home-convalescing Dad. Like Alison's description of the greys in her care, he's already shown considerable resilience after critical illness and a determination to 'keep on fighting till his last breath' and I'm hoping he'll be able to move calmly through this next phase of his rehabilitation.

In the meantime, Alison[1] sends me on my way with the promise that, in a month or so, if they continue to make good progress, the two Pembrokeshire seals in the seventeenth-century barn will be able to follow me home.

EPILOGUE

Adult Returners

In the fields behind the coast path, late haymaking is underway. Some meadows are already mown, with sun-warmed windrows ready for baling. Another is in the process of being cut and a couple of buzzards are circling, alert to the disturbance of small mammalian prey.

It's late August and the smell of the mown crop blends with the honeyed scent of meadowsweet as Hooper and I wend our way above the bay at Aber Felin, passing clumps of ragwort reeling with gatekeeper butterflies. As I step off the path for a view over the first beach, I gatecrash a common lizard's sunbath, sending it darting into a patch of bracken. There's an extra-low tide today with unexpected rocks exposed and thirteen seals, the most I've seen here since spring, flopped on top of them. Several seals are lolling, too, on a bed of kelp that's likewise being revealed by the tide, flippers and fronds wafting and mingling. While some people might rail at reaching the tail end of summer, I love this time of year, with pregnant cows, who've chosen the bay to be this season's birthing place, starting to gather. Knowing that many females exhibit site fidelity, going back to the same place to give birth each year, while others opt for a different location, I can assume I'm viewing a mix of newcomers and returners.

In the months between the end of the moult and this prologue to the pupping season, while the seals have mostly been away at their summer feeding grounds, Dad's been continuing his recuperation. His period of convalescence at the care home has evolved into a long-term stay that's likely to become permanent. Remarkably, after years of asserting his desire to live independently and resisting offers of help, he's admitted that he's now quite enjoying not having to cook, clean, shop and carry out all the other essential tasks that are involved in running a house. The regular monitoring of his medical conditions also brings me much comfort: having fretted my way through the last couple of years, I no longer have to worry that he'll suffer a dementia-related or cardiovascular crisis while he's on his own with no one on hand to help him. However, though my anxiety's lessened, his confinement to a care home has dredged up a sludge of guilt in spite of the fact that he reached the decision to stay there himself. Like Flipper and Ray and the other permanent residents of the Cornish Seal Sanctuary with ongoing health issues whom I so relished encountering at the start of my seal journey, he's been rescued and rehabilitated but the third stage – release back into the wild – hasn't been realised, a loss about which I feel deeply sad.

Hooper's nose nudges the back of my thigh, hinting that we've lingered here long enough and he's ready to walk on. A mixed flock of goldfinches and linnets chitters and flitters between gorse and barbed wire fence as we pass above beach number two. Though I manage to get him to step around a fox moth caterpillar that's making its hairy, undulating way across the path, I opt against coaxing him onto the ledge overlooking the third beach and we continue to watch from the cliff top instead.

Just offshore, two more pregnant cows have draped themselves on a rock around which a bull seal's circling, another sign that the breeding season is imminent. One of the cows pays him no attention and carries on dozing on her side. The other is much more bothered by his presence and lets out a succession of disgruntled vocalisations – a snort, a howl, a phlegmy, clearing-of-the-throat-type sound – whenever he shows signs of wanting to heave himself out of the water to join them. The first cow continues to ignore him, rousing only to rub her nose with her front flipper. As she lowers her head to her pillow of rock again, my attention and binoculars shift more fully from the bull's shenanigans to the pattern of her fur, the sequence of blotches that punctuates the narrative of the left side of her neck. I need no mammoth photo ID catalogue to tell me that this is a pattern – three inverted commas and a full stop – that I recognise. When she raises her head again and a scar just above her right eye is also revealed, I'm able to unequivocally confirm her as the cow who was spotlit by the low evening sun when I picnicked here with Hooper last summer. At long last, I've managed to notch up my first Aber Felin fur pattern match.

In addition to my delight at this first repeat sighting, each visit to Aber Felin is now mightily enriched by all I've learnt on my travels to other seal colonies and by the insights I've gleaned from seal specialists en route. One of my original impulses for undertaking my journey was to gain a more acute appreciation of the threats grey seals are suffering and every site with which I've become familiar seems to have offered a fresh perspective on these challenges. Both at sea and on land, in the course of conducting their daily and seasonal routines, seals experience physiological stress – by diving on a single held breath, for example, and by fasting for weeks at a time during the pupping season – and the human-induced stresses to which they're exposed in both habitats add a damaging load. While some colonies currently appear to be thriving, enough warnings were sounded during the course of my travels to suggest that human-generated threats could have a serious, long-term, cumulative effect on grey seals' well-being and breeding success.

Yet in spite of the fact that the risk of loss, both ecological and personal, has been a recurring theme over the past couple of years of my seal journey, I've simultaneously come across a whole host of positives. Animal care workers nursing seals through grievous marine litter injuries and facilitating their release back into the wild.

Scientists conducting pioneering research into the metabolic implications of toxic pollutants. Volunteers tirelessly wardening at pupping sites and lobbying for better protection of haul-outs. All of them drawn to defending a species in which so many human interests – cultural, political, economic and environmental – reside and collide.

From a personal point of view, spending time in the company of seals has repeatedly soothed and fortified me as I've struggled to reconcile myself both to Dad's diminished cognition and his life-threatening heart condition. Moreover, now that he's ensconced in the care home, seals are one of the means by which we can continue to share our pleasure in the natural world. While we may no longer be able to spot birds and other wildlife on walks together, we can still connect through nature-themed conversation. I show him photos and videos from my visits to seal sites and we talk about the rhythms of their lives. 'They have their pups in the autumn and not in the spring?' he says every time. 'Are you sure?'

I keep him well supplied with books and whenever I visit, he greets me with a new discovery. 'Did you know that trees can talk to each other through their roots underground?' 'Have a guess which bird's got the best hearing.' 'Bet you never knew we've got more than – hang on, how many is it? – six hundred species of spider.' In a poignant throwback to my childhood when he read aloud, easing me into sleep to relieve my night terrors, he often regales me with whole chapters, with Hooper lying between us, his tail beating against the carpet in response to the changing inflections of Dad's voice. The canine and pinniped families may well have diverged around fifty million years ago, but, just as being in the presence of seals has consistently been a source of solace to me, so Hooper has never failed to give Dad joyful and steadfast comfort.

Now, instead of a thumping tail, I feel Hooper's nose giving my thigh another nudge. The sun has temporarily retreated and a cluster of plump grey clouds is hauled out along the horizon. The bull has also retreated, swimming off to rehearse his beachmaster skills with the cows on the kelp bed.

Picking up on Hooper's cue, I retreat too, cradling, as we make for home, the memory of my twice-seen seal and her unborn pup.

For now, at least, sleeping undisturbed.

Notes

2. Sanctuary

1 While most sources continue to assert that there are thirty-three species of pinniped, recent genetic analysis is suggesting that this figure needs to be updated, with the Galápagos sea lion, so long regarded as a subspecies of the California sea lion, being reclassified as a species in its own right. Another fluctuation in species numbers in recent decades was triggered by the tragic extinction of the Caribbean monk seal due to unrestrained hunting, and overfishing of its food. Because several other pinnipeds, including the Mediterranean monk seal, are severely endangered, it's also possible that, in spite of conservation efforts, there could be a decrease in the number of recognised species in the future.

2 This challenge is now thought to have lessened as there's been a change in ownership since I visited the Cornish Seal Sanctuary: Merlin Entertainments has gifted the sanctuary to its partner marine life charity, the Sea Life Trust. As a result, the sanctuary is now said to be able to retain more of its visitor revenue and focus more fully on its core seal rehab and welfare efforts. https://www.sealsanctuary. co.uk/html/pressrelease20180319css.html

3. Seal HQ

1 For further information about Cornwall Seal Group Research Trust's groundbreaking conservation, education and campaigning work, which has advanced apace since my meeting with Sue Sayer, see https://www.cornwallsealgroup.co.uk/

   CSGRT has also recently adopted, and registered with the Charity Commission, a second working name – Seal Research Trust – to better reflect its contribution to seal conservation at a national level.

2 Apart from the new ownership, there have been a few other changes at the Cornish Seal Sanctuary since my most recent visit. Sadly, three of the elderly grey seal residents – Flipper, Snoopy and Lizzie – have died at the venerable ages of thirty-six, thirty-nine and forty-one, as has the younger Badger, while Dan Jarvis has left the sanctuary to take up a new role as Welfare Development and Field Support Officer at British Divers Marine Life Rescue.

4. Dye Hard

1 In an 1897 lecture to Cardiff Naturalists Society, *Notes on a Visit to Skomer*, J. J. Neale wrote that 'On two or three occasions we saw seals . . . The so-called gentlemen take a delight in shooting them although they know full well that if shot while in the water the bodies at once sink' while in *Cliffs of Freedom: The Story of Skomer Island and the Last Man to Farm it* (Gomerian Press, 1961), historian Roscoe Howells observed that 'Only a chance seal was seen. Fishermen shot them when they could.'

2 Bee Büche and Ed Stubbings no longer have to deal with bull seal/ Zodiac issues as, after working for six years as joint wardens on Skomer, they've relocated.

5. ExtractCompare

1 Though the tidal turbine's problems haven't been resolved since my day on Ramsey and the unit hasn't generated any more electricity,

valuable acoustic monitoring information with respect to the movement of cetaceans around the turbine was garnered during the short period of its operation. Some of the results of this research are summarised at: http://www.smruconsulting.com/?p=12947

There's also been some indication that Ramsey Sound could be revitalised as a tidal energy site as part of TIGER, an ambitious Anglo-French Tidal Stream Industry Energiser Project. In the first instance, the plan is for the turbine to be removed and the reasons for its failure analysed. https://www.theengineer.co.uk/tiger-tidal-stream-project/

2 A follow-up to this preliminary collision trials study, again featuring a turbine blade replica and grey seal carcasses, has also now been published. Although none of the nineteen carcasses manifested lumbar or cervical spine injuries, acute damage to the thoracic area was observed, in contrast to the no-skeletal-trauma findings of the earlier investigation. Speed of impact had a bearing on the severity of injury, as did the body condition of each of the deceased seals – the thicker the blubber the less extreme the trauma. https://besjournals.onlinelibrary.wiley.com/doi/10.1111/1365-2664.13388

3 After almost fourteen years as Ramsey's co-warden, Lisa Morgan has left the island to work as Head of Islands and Marine with the Wildlife Trust of South and West Wales.

8. The Dawn Song of Middle Eye

1 'Crime Writing Capital of the UK', *BBC One, Inside Out – North West* (16 September 2002): http://www.bbc.co.uk/insideout/northwest/series1/crime-writers.shtml

9. Action Man

1 The official Isle of Man government website: https://www.gov.im/about-the-government/departments/cabinet-office/external-relations/constitution/

2 'Jeremy Clarkson's anger at ramblers who disturb his peace at lighthouse home', *Mail Online* (25 July 2009): https://www.dailymail.

co.uk/news/article-1201907/Jeremy-Clarksons-anger-ramblers-disturb-peace-lighthouse-home.html

3   'Jeremy Clarkson loses footpath battle', *The Telegraph* (11 May 2010): https://www.telegraph.co.uk/news/7712196/Jeremy-Clarkson-loses-footpath-battle.html

4   Ibid.

10. Learning to Speak Seal

1   *Sunday Mail* (18 June 2017)

2   'Watch seal pups being soothed by classical flute player's Christmas tunes at Cornish Seal Sanctuary', *Cornwall Live* (20 December 2017): https://www.cornwalllive.com/news/cornwall-news/watch-seal-pups-being-soothed-954976 and 'Seals serenaded by classical flautist in Cornwall', *BBC News* (21 December 2017): https://www.bbc.co.uk/news/av/uk-england-cornwall-42446045

3   *Daily Record* (20 June 2019): https://www.dailyrecord.co.uk/news/weird-news/seals-sing-star-wars-tune-16549265

4   'Grey seals can learn to sing like humans, says study', *BBC News* (20 June 2019): https://www.bbc.co.uk/news/uk-scotland-edinburgh-east-fife-48706251

12. Skins of Silver, Chains of Gold

1   The Conservation of Seals Act now applies only to England and Wales as it was superseded in Scotland by the Marine (Scotland) Act 2010.

2   In 2021, significant amendments to this legislation were implemented – chapter 13 and, specifically, note 5 on p. 349 offer further information and context.

3   'Seals costing fishing industry £25 million a year', *Daily Telegraph* (13 February 1978), quoted on p. 17 of *Seal Cull: the Grey Seal Controversy* by John Lister-Kaye (London: Penguin, 1979)

13. The Scramble for Salmon

1 'Tourist couple's shock over seal cull', *The Herald* (21 August 2012): https://www.heraldscotland.com/news/13070028.tourist-couples-shock-over-seal-cull/

2 Ibid.

3 'Seal killing quota labelled "pointless"', *The Herald* (21 April 2013): https://www.heraldscotland.com/news/13101158.seal-killing-quota-labelled-pointless/

4 Ibid.

5 Seal Defence Campaign was Sea Shepherd UK's name for the crusade and, because this was the term that appeared most frequently in the media, I've also opted to adopt it. The Hunt Saboteurs Association, however, preferred to use a different name – Seal Guardian Campaign.

6 Quoted on p. 112 of *The Devil and the Deep Blue Sea: An Investigation into the Scapegoating of Canada's Grey Seal* by Linda Pannozzo (Black Point, Nova Scotia: Fernwood Publishing, 2013)

7 Concurrent with his Seal Protection Action Group role, Andy Ottaway has been the director of Campaign Whale, an organisation working to safeguard dolphins, whales and their habitats, and is therefore uniquely placed to comment on the disparity in concern typically shown towards seals and cetaceans.

8 '1,500 seals shot in four years by Scottish salmon industry', Seal Protection Action Group blog post (1 October 2015): https://www.sealaction.org/1500-seals-shot-in-four-years-by-scottish-salmon-industry

9 https://twitter.com/DavidBowles21/status/1039493252091703296

10 Andy Ottaway's prediction was accurate: money does indeed talk. In June 2020, a repeal of the provision to grant licences for the killing of seals was announced by the Scottish parliament. While officials have implied that the decision was driven by the intention to improve seals' welfare, it's widely accepted that it was made in response to the new US regulations preventing Scotland from continuing its lucrative exports should fish farms keep controlling seals by lethal means. Seven months later, at the end of January 2021, a ban on seal shooting in Scotland finally came into force.

The UK government swiftly followed suit, with amendments, from March 2021, to the Conservation of Seals Act and the Wildlife (Northern Ireland) Order likewise making it illegal to intentionally kill or injure seals in England, Wales, Northern Ireland and their territorial waters.

14. Stampede

1 Scottish Natural Heritage underwent a rebrand in 2020 and is now known as NatureScot.

16. The Hubris of Scuba

1 'Lovable Seal Hugs Diver', *Right This Minute! – The Viral Videos Show* (5 October 2018): https://www.youtube.com/watch?v=zbkp8QoJC8k

   'Wild Grey Seal Gives Doctor Loving Embrace Whilst Diving Off Coast', *Caters Clips* (4 October 2018): https://www.youtube.com/watch?v=xCqyYtDZUJ4

   'Posing for a seal-fie! Diver makes friends with a seal who smothers him in "kisses" in adorable footage', *Mail Online* (5 October 2018): https://www.dailymail.co.uk/news/article-6244355/Diver-makes-friends-seal-smothers-kisses-adorable-footage.html

2 On his YouTube channel, Ben Burville has latterly addressed the harassment risk issue, posting several videos with an accompanying warning that divers should refrain from touching marine life due to the potential hazards, a justification of his own underwater interactions with seals, and an insistence that they should never be disturbed, especially when hauled out on land. See, for example: https://www.youtube.com/watch?v=R73WrYJkw6o and https://www.youtube.com/watch?v=_9xGW7vDRFo

17. Industrial Strength

1 St Mary's Seal Watch recently changed its name to St Mary's Island Wildlife Conservation Society, a move that Sal Bennett's comment and avowed interest in the wider island environment seems to anticipate.

19. Sonic Boom

1 There has been a change in personnel at Donna Nook since my visit, with both Rob Lidstone-Scott and Lizzie Lemon leaving the reserve and their posts as wardens with Lincolnshire Wildlife Trust.

20. The Count

1 Since I joined him for the pup count, Ajay Tegala has moved on from his role as National Trust ranger at Blakeney Point and now combines his work at Wicken Fen Nature Reserve in Cambridgeshire, again as National Trust ranger, with wildlife TV presenting.

The method by which the Blakeney pups are counted has also markedly changed. Instead of rangers moving through the whole colony, attempting to record every newborn, they're now documenting births in just one area, from which, over time, an assessment of the size of the entire colony can be made.

21. Birth and Rescue

1 All the numbers of pup births that I mention in the various east coast breeding colonies pertain to, and were accurate at, the time of my visits. Up-to-date figures are usually shared every pupping season on the social media feeds and/or blogs of the organisations that are responsible for managing/monitoring each site, or in an end-of-season press release. Some useful links are as follows:

https://www.facebook.com/isleofmayNNR and https://twitter.com/SteelySeabirder (Isle of May);

https://www.facebook.com/DonnaNookWarden and https://twitter.com/DonnaNookWarden (Donna Nook);

https://www.facebook.com/FriendsofHorseySeals and https://twitter.com/HorseySeals (Horsey).

2 Disturbance incidents on the beach at Winterton-on-Sea during the pupping season have continued to escalate, with several whitecoats shockingly losing their lives. A fence has therefore finally been erected

by Natural England to afford the seals greater protection and create a viewing corridor for visitors. https://www.wintertononsea.co.uk/blog/Entries/2020/10/protecting-the-seals.html

23. Moult

1   In 2021, after more than three decades of dedication to the animal welfare cause at RSPCA East Winch, Alison Charles took early retirement, and the centre's vital seal rescue and rehab work is now continuing under a new manager.

Bibliography

The following list is not exhaustive but includes the books, articles, reports and web resources to which I've either made reference in the text or which I've found especially helpful and inspiring.

Web addresses often change: those that appear below were accurate when this book went to press.

Alexander, Rosanne, *Selkie* (London: André Deutsch, 1991)

Alexander, Rosanne, *Waterfalls of Stars* (Bridgend: Seren, 2017)

Allan, James, *General Guide to the Isle of May* (Anstruther: Tervor Publishing, 2000)

Anderson, Sheila, *The Grey Seal* (Aylesbury: Shire Publications, 1988)

Andersson, M. H., Andersson, S., Ahlsén, J. et al., *A framework for regulating underwater noise during pile driving,* a technical Vindval report, Swedish Environmental Protection Agency (2016)

Baines, M. E., Earl, S. J., Pierpoint, C. J. L. & Poole J., *The West Wales Grey Seal Census*, Countryside Council for Wales Contract Science Report 131 (1995)

Barkham, Patrick, 'Waves of Destruction', *Guardian* (17 April 2008): https://www.theguardian.com/environment/2008/apr/17/flooding.climatechange

Barkham, Patrick, *Islander: A Journey Around Our Archipelago* (London: Granta Books, 2017)

Barnes, Geoff, 'Hilbre Island to be without a custodian as council struggles to find lone ranger', *Wirral Globe* (17 January 2011): https://www.wirralglobe.co.uk/news/8794988.hilbre-island-to-be-without-a-custodian-as-council-struggles-to-find-lone-ranger/

Bathurst, Bella, *The Lighthouse Stevensons* (London: HarperCollins, 1999)

BBC News, 'Grey seal spotted in Bewdley and Stourport-on-Severn' (3 January 2013): https://www.bbc.co.uk/news/uk-england-hereford-worcester-20897600

BBC News, 'Norfolk grey seal twin pups R2-D2 and C-3PO "a world first"' (7 March 2016): https://www.bbc.co.uk/news/uk-england-norfolk-35745612

Bellman, K., Bennett, S., James-Hussey, A. et al., *PLEASE DO NOT DISTURB! The growing threat of seal disturbance in the United Kingdom: Case studies from around the British coast,* report commissioned by The Seal Alliance (2019)

Best Fishes: https://www.bestfishes.org.uk/

Bishop, A. M., Lidstone-Scott, R., Pomeroy, P. & Twiss, S., 'Body slap: An innovative aggressive display by breeding male gray seals (*Halichoerus grypus*)', *Marine Mammal Science*, vol. 30, no. 2, pp. 579–593 (2014)

Bishop, A. M., Denton, P., Pomeroy, P. & Twiss, S., 'Good vibrations by the beach boys: magnitude of substrate vibrations is a reliable indicator of male grey seal size', *Animal Behaviour*, 100, pp. 74–82 (2015)

Bodin, Madeline, 'Mystery of the Corkscrew Seals', *bioGraphic* (24 May 2016): https://www.biographic.com/mystery-of-the-corkscrew-seals/

Bolt, H. E., Harvey, P. E., Mandleberg, L. & Foote, A. D., 'Occurrence of killer whales in Scottish inshore waters: temporal and spatial patterns relative to the distribution of declining harbour seal populations', *Aquatic Conservation: Marine and Freshwater Ecosystems*, vol. 6, issue 6, pp. 671–675 (2009)

Bond, Ian, *Tees Seals Research Programme Monitoring Report No. 31,* produced by the Industry Nature Conservation Association (2019)

Bowen, D., '*Halichoerus grypus,* grey seal', *The IUCN Red List of Threatened Species* (2016)

Bradfield, Elizabeth, *The Haul Out: Considering Seals and Other Items*

Bibliography

Ashore, Mostly on Cape Cod, blog (2013–2018): https://thehaulout.wordpress.com

Brand, John, *A brief description of Orkney, Zetland, Pightland-Firth & Caithness* (1701), Oxford Text Archive: http://hdl.handle.net/20.500.12024/K112620.000

Brasseur, S. M. J. M., van Polanen Petel, T. D., Gerrodette, T. et al., 'Rapid recovery of Dutch gray seal colonies fueled by immigration', *Marine Mammal Science,* vol. 31, issue 2 (2015)

Bridgeman, H. & Wylor-Owen, J., *Calf of Man Seal Surveys Autumn Report* for Manx Wildlife Trust (2017)

Bristow, Su, *Sealskin* (London: Orenda Books, 2017)

Brownlow, A., Onoufriou, J., Bishop, et al., 'Corkscrew seals: grey seal (*Halichoerus grypus*) infanticide and cannibalism may indicate the cause of spiral lacerations in seals', *PLoS ONE,* vol. 11, no. 6, pp. 1–14 (2016)

Büche, B. & Stubbings, E., *Grey Seal Breeding Census Skomer Island,* Natural Resources Wales annual reports (2013–2019)

Burville, Ben, YouTube channel (2006–present): https://www.youtube.com/channel/UCIEvGGGTGbmr_v97PNaadCg

Burville, Ben, Twitter profile @Sealdiver (2009–present): https://twitter.com/Sealdiver

Butterworth, Andy (ed.), *Marine Mammal Welfare: Human Induced Change in the Marine Environment and its Impacts on Marine Mammal Welfare* (Cham: Springer, 2017)

Carter, M. I. D., Russell, D. J. F., Embling, C. B. et al., 'Intrinsic and extrinsic factors drive ontogeny of early-life at-sea behaviour in a marine top predator', *Scientific Reports,* vol. 7, article no. 15505 (2017)

Clarkson, Ewan, *Halic, the Story of a Grey Seal* (London: Arrow Books, 1972)

Coram, A., Gordon, J., Thompson, D. & Northridge, S., *Evaluating and assessing the relative effectiveness of non-lethal measures, including Acoustic Deterrent Devices, on marine mammals,* a Scottish Government-commissioned report (2014)

Corey Dunne, Brenda, *Skin* (Fortunate Frog Fiction, 2016)

Cornwall Seal Group Research Trust, blog featuring latest news (2013–present): https://www.cornwallsealgroup.co.uk/news/

Cornwell, Betsy, *Tides* (New York: Houghton Mifflin Harcourt, 2013)

Davies, Brian, *Red Ice: My Fight to Save the Seals* (London: Methuen, 1989)

Davis, Randall W., *Marine Mammals, Adaptations for an Aquatic Life* (Cham: Springer, 2019)

De Winter, Gunnar, 'Do Seals Get Alzheimer's Disease?', *Predict* (2 February 2021): https://medium.com/predict/do-seals-get-alzheimers-disease-8b053256f599

Dickensen, Victoria, *Seal* (London: Reaktion Books, 2016)

Doherty, Berlie, *Daughter of the Sea* (London: Andersen Press, 2008)

Doolittle, Emily, *Seal Songs* [music score], with Gaelic texts by Rody Gorman, for the Voice Factory Choir and the Paragon Ensemble (2011)

Doolittle, Emily, *Conversation* [music score], based on poetry by Eleonore Schönmaier, for the St Andrews New Music Ensemble (2018)

Duck, Callan, *Seals*, Scottish Natural Heritage (2007)

Dunsford, Cathie, *Song of the Selkies* (Melbourne: Spinifex Press, 2001)

Edwards, Rob, investigative journalism on salmon farming, *The Ferret* (2016–present): https://theferret.scot/tag/salmon/

Express & Star, 'Campaign is on to save River Severn seal Keith' (14 January 2015): https://www.expressandstar.com/news/2014/01/15/campaign-is-on-to-save-river-severn-seal-keith/

Farre, Rowena, *Seal Morning* (Edinburgh: Birlinn, 2008)

Fink, Sheryl, 'Could 2020 finally bring an end to the commercial seal hunt in Canada?' (5 June 2020): https://www.ifaw.org/uk/journal/end-commercial-seal-hunt-canada-2020

The Fish Site, 'Predator nets reduce seal cull by a third' (February 2019): https://thefishsite.com/articles/predator-nets-reduce-seal-cull-by-a-third

The Fish Site, 'Scots to ban seal shooting on fish farms' (18 June 2020): https://thefishsite.com/articles/scots-to-ban-seal-shooting-on-fish-farms

Flint, Sue, *Let the Seals Live!* (Sandwick: The Thule Press, 1979)

Fowlis, Julie, 'An Ròn' ('The Seal') and 'Ann An Caolas Od Odrum' ('In the Narrows of Od Odrum') on *Gach sgeul – Every story* [CD] (Machair Records, 2014)

Fowlis, Julie, 'Òran an Ròin' ('The Seal's Song') on *Alterum* [CD] (Machair Records, 2017)

Fraser Darling, Frank, *A Naturalist on Rona* (Oxford: Clarendon Press, 1939)

Bibliography

Friends of Hilbre, Seal Watch Records (2007–2017)

Gill, Victoria, 'Seal whiskers sense faraway fish', *BBC News* (11 June 2010): https://www.bbc.co.uk/news/10287564

Green, Jenny, 'Stopping the Scottish Seal Slaughter', *Earth First! Journal* (2016)

Hall, A. J., Jepson, P. D., Goodman, S. J. & Härkönen, T., 'Phocine distemper virus in the North and European seas – Data and models, nature and nurture', *Biological Conservation*, vol. 131, pp. 221–229 (2006)

Hammond, P. S. & Wilson L. J., *Grey Seal Diet Composition and Prey Consumption*, Scottish Marine and Freshwater Science Report, vol. 7, no. 20 (2016)

Härkönen, T., Dietz, R., Reijnders, P. et al., 'A review of the 1988 and 2002 phocine distemper virus epidemics in European harbour seals', *Diseases of Aquatic Organisms*, vol. 68, pp. 115–130 (2006)

Hastie, G. D., Russell, D. J. F., McConnell, B. et al., 'Sound exposure in harbour seals during the installation of an offshore wind farm: predictions of auditory damage', *Journal of Applied Ecology*, vol. 52, issue 3, pp. 631–640 (2015)

Hersey, Lu, *Deep Water* (London: Usborne Publishing, 2015)

Hickling, Grace, *Grey Seals and the Farne Islands* (London: Routledge and Kegan Paul, 1962)

Hocking, D. P., Marx, F. G., Sattler, R. et al., 'Clawed forelimbs allow northern seals to eat like their ancient ancestors', *Royal Society Open Science*, vol. 5, issue 4 (2018)

Hocking, D. P., Burville, B., Parker, W. M. G. et al., 'Percussive underwater signaling in wild gray seals', *Marine Mammal Science*, vol. 36, issue 2 (2020)

Horning, M., Andrews, R. D., Bishop, A. M. et al., 'Best practice recommendations for the use of external telemetry devices on pinnipeds', *Animal Biotelemetry*, vol. 7, issue 20 (2019)

Howells, Roscoe, *Cliffs of Freedom: the story of Skomer Island and the last man to farm it* (Llandysul: Gomerian Press, 1961)

Hunter, Mollie, *A Stranger Came Ashore* (Edinburgh: Kelpies, 2012)

Hurrell, H. G., *Atlanta, My Seal* (London: William Kimber, 1963)

International Council for the Exploration of the Sea, *Report of the Workshop on Predator-prey Interactions between Grey Seals and other marine mammals* (2017)

Jones, Ken, *Seal Doctor* (Glasgow: Fontana/Collins, 1986)

Kelly, John, 'From cold waters to Cold War: The story of a Navy seal', *The Washington Post* (26 June 2012): https://www.washingtonpost.com/local/from-cold-waters-to-cold-war-the-story-of-a-navy-seal/2012/06/26/gJQA4kA84V_story.html

Kiely, O., Lidgard, D., McKibben, M. et al., *Grey Seals: Status and Monitoring in the Irish and Celtic Seas,* Maritime Ireland/Wales INTERREG Report no. 3 (2000)

King, Rachael, *Red Rocks* (Auckland: Random House, 2012)

Kingshill, Sophia, *Mermaids* (Dorset: Little Toller, 2015)

Kingshill, Sophia & Westwood, Jennifer, *The Fabled Coast* (London: Arrow Books, 2014)

Lambert, R. A., 'The Grey Seal in Britain: A Twentieth Century History of a Nature Conservation Success', *Environment and History,* vol. 8, no. 4, pp. 449–474 (2002)

Lambton, Lucinda, *Temples of Convenience and Chambers of Delight* (London: Pavilion Books, 1997)

Lanagan, Margo, *The Brides of Rollrock Island* (Oxford: David Fickling Books, 2013)

Lawson, B. & Jepson, P., *UK Phocine Distemper Virus Epizootic Investigation Report 2002/2003*, produced by the Institute of Zoology at the Zoological Society of London for the Department for Environment, Food and Rural Affairs (2003)

Lennon, Joan, *Silver Skin* (Edinburgh: Birlinn, 2015)

Lister-Kaye, John, *Seal Cull: the Grey Seal Controversy* (London: Penguin, 1979)

Lockley, R. M., *The Seals and the Curragh* (London: Dent, 1954)

Lockley, R. M., *Seal Woman* (London: Rex Collings, 1974)

Lockley, R. M., *Grey Seal, Common Seal* (London: White Lion Publishers, 1977)

Lockley, R. M., *Dream Island* (Dorset: Little Toller, 2016)

Macdonald, Michael, 'Government looking at plan to revive seal penis sales', *The Globe and Mail* (7 June 2015): https://www.theglobeandmail.

com/news/national/government-looking-at-plan-to-revive-seal-penis-sales/article24835392/

Mâche, François-Bernard, *Music, Myth and Nature, or The Dolphins of Arion* (London: Routledge, 1993)

Malinka, C. E., Gillespie, D. M., Macaulay, J. D. J. et al., 'First in-situ passive acoustic monitoring for marine mammals during operation of a tidal turbine in Ramsey Sound, Wales', *Marine Ecology Progress Series*, 590: 245–266 (2018)

Mara: the Seal Wife [film], directed by Uisdean Murray, UK: Lily Islands Films (2021): https://www.mara.film/our-journey

Marine Scotland, 'Guidance on the Offence of Harassment at Seal Haul-Out Sites' (2014)

Marshall, Michael, 'Seals remember what they just did – but only for about 18 seconds', *New Scientist* (3 July 2019): https://www.newscientist.com/article/2208530-seals-remember-what-they-just-did-but-only-for-about-18-seconds/

McConnell, B. J., Fedak, M. A., Lovell, P. & Hammond, P. S., 'Movements and foraging areas of grey seals in the North Sea', *Journal of Applied Ecology*, vol. 36, issue 4, pp. 573–590 (1999)

McCulloch, Susanne, 'The Vocal Behaviour of the Grey Seal (*Halichoerus grypus*)', PhD thesis, University of St Andrews (2000)

McDonald, Patrick, *A Collection of Highland Vocal Airs* (Isle of Skye: Taigh na Teud, 2000)

McEntire, N. C., 'Supernatural Beings in the Far North: Folklore, Folk Belief and the Selkie', *Scottish Studies*, vol. 38, pp. 120–143 (2018)

Middleton, Fiona, *Seal* (Edinburgh: Mainstream Publishing, 1995)

Molloy, Pat, *Operation Seal Bay* (London: Hodder & Stoughton, 1991)

Morgan, Greg, 'Ophelia Hits Hard', Ramsey Island blog (18 October, 2017): https://community.rspb.org.uk/placestovisit/ramseyisland/b/ramseyisland-blog/posts/ophelia-hits-hard

Muir, Tom, *The Mermaid Bride and Other Orkney Folk Tales* (Kirkwall: Kirkwall Press, 1998)

National Trust, Blakeney Grey Seal Pup Counts (1988; 1997–1999; 2000–present)

Neale, J. J., *Notes on a Visit to Skomer*, lecture to Cardiff Naturalists Society (1897)

Nelms, S. E., Galloway, T. S., Godley, B. J. et al., 'Investigating microplastic trophic transfer in marine top predators', *Environmental Pollution*, vol. 238, pp. 999–1007 (2018)

Nelms, S. E., Barnett, J., Brownlow, A. et al., 'Microplastics in marine mammals stranded around the British coast: ubiquitous but transitory?', *Scientific Reports* 9, article no. 1075 (2019)

Norris, Samuel, *Douglas, the Naples of the North* (1904)

Ondine [film], directed by Neil Jordan, Ireland: Octagon Films (2009)

Onoufriou, J., Brownlow, A., Moss, S. et al., 'Empirical determination of severe trauma in seals from collisions with tidal turbine blades', *Journal of Applied Ecology*, vol. 56, issue 7, pp. 1712–1724 (2019)

Orams, M. B., 'Feeding wildlife as a tourism attraction: a review of issues and impacts', *Tourism Management*, 23, pp. 281–293 (2002)

Palmer, Roy (ed.), *A Book of British Ballads* (Somerset: Llanerch Press, 1998)

Pannozzo, Linda, *The Devil and the Deep Blue Sea: An Investigation into the Scapegoating of Canada's Grey Seal* (Black Point, Nova Scotia: Fernwood Publishing, 2013)

Paro Therapeutic Robot: https://www.paroseal.co.uk/

Payne, Andrew & Shields, Mark (eds), *Marine Renewable Energy Technology and Environmental Interactions* (Dordrecht: Springer, 2014)

Pearson, R. H., *A Seal Flies By* (London: Rupert Hart-Davies, 1959)

Pinkola Estés, Clarissa, *Women Who Run With the Wolves* (London: Rider, 1992)

Pliny the Elder, *Natural History*, selected and translated by John F. Healy (London: Penguin, 1991)

Pomeroy, P. P., Twiss, S. D. & Redman P., 'Philopatry, Site Fidelity and Local Kin Associations within Grey Seal Breeding Colonies', *Ethology: International Journal of Behavioural Biology*, vol. 106, issue 10, pp. 899–919 (2000)

Ridgway, S. H., Harrison, R. J. & Joyce, P. L., 'Sleep and Cardiac Rhythm in the Gray Seal', *Science*, vol. 187, no. 4176, pp. 553–555 (1975)

Riedman, Marianne, *The Pinnipeds* (Berkeley: University of California Press, 1990)

Riordan, Julie, 'Seals and Storms', Skomer Island blog (22 October 2017): https://skomerisland.blogspot.com/2017/10/seals-and-storms.html

Robinson, K. J., Hall, A. J., Debier, C. et al., 'Persistent organic pollutant burden, experimental POP exposure and tissue properties affect metabolic profiles of blubber from grey seal pups', *Environmental Science and Technology*, vol. 52, no. 22, pp. 13523–13534 (2018)

Robinson, K. J., Hall, A. J., Scholl, G. et al., 'Investigating decadal changes in persistent organic pollutants in Scottish grey seal pups', *Aquatic Conservation*, vol. 29, issue S1, *Supplement: Sea Mammal Research Unit 40th Anniversary: Advances in Marine Mammal Science Informing Policy*, pp. 86–100 (2019)

Robinson, Kelly, *Kelly Robinson Science: Post Doctoral Marine Mammal Research in Behaviour and Physiology*, blog (2016–2019): https://kellyrobinsonscience.wordpress.com/blog/

Roman, J. & McCarthy, J. J., 'The Whale Pump: Marine Mammals Enhance Primary Productivity in a Coastal Basin', *PLoS ONE*, vol. 5, no. 10: e13255 (2010)

Russell, D. J. F., Brasseur, S. M. J. M., Thompson, D. at al., 'Marine mammals trace anthropogenic structures at sea', *Current Biology*, vol. 24, issue 14, pp. R638–R639 (2014)

Russell, D. J. F. & McConnell, B., *Seal at-sea distribution, movements and behaviour*, report for the Department of Energy and Climate Change (2014)

Russell, D. J. F., *Movements of grey seal that haul out on the UK coast of the southern North Sea*, report for the Department of Energy and Climate Change (2016)

Sackville, Amy, *Orkney* (London: Granta Books, 2013)

Sayer, Sue, *Seal Secrets: Cornwall and the Isles of Scilly* (Penzance: Alison Hodge Publishers, 2012)

Schönmaier, Eleonore, *Treading Fast Rivers* (Ottawa: Carleton University Press, 1999)

Schop, J., Aarts, G., Kirkwood R. et al., 'Onset and duration of gray seal (*Halichoerus grypus*) molt in the Wadden Sea, and the role of environmental conditions', *Marine Mammal Science*, vol. 33, issue 3, pp. 830–846 (2017)

Sea Shepherd UK, 'Sea Shepherd UK's Reply to Usan Fisheries Ltd, trading as the Scottish Wild Salmon Company', *Sea Shepherd News* (2014): https://www.seashepherd.org.uk/news-and-commentary/news/sea-shepherd-uk-s-reply-to-usan-fisheries-ltd-trading-as-the-scottish-wild-salmon-company.html

Seal Conservation Society, 'Grey Seal (*Halichoerus grypus*)' (2011): https://www.pinnipeds.org/seal-information/species-information-pages/the-phocid-seals/grey-seal

Seal Protection Action Group, blog (2008–2018): http://www.sealaction.org/news

Selkie [film], directed by Donald Crombie, Australia: Bluestone Pictures (2000)

Selman, R. G. & Cherrill, A. J., 'The lesser mottled grasshopper, *Stenobothrus stigmaticus*: lessons from habitat management at its only site in the British Isles', *Journal of Orthoptera Research*, vol. 27, issue 1, pp. 83–89 (2018)

Sharpe, Chris, *Report on a survey of Grey Seals around the Manx coast, undertaken from April 2006 to March 2007*, produced by Manx Bird Atlas for the Department of Agriculture, Fisheries & Forestry (2007)

Smithsonian's National Zoo Press Office, 'Elderly Grey Seal Dies at Smithsonian's National Zoo' (25 June 2012 & 10 November 2016): https://www.si.edu/newsdesk/releases/elderly-gray-seal-dies-smithsonian-s-national-zoo and https://www.si.edu/newsdesk/releases/elderly-gray-seal-dies-smithsonian-s-national-zoo-0

Song of the Sea [animated film], directed by Tomm Moore, Ireland: Cartoon Saloon (2014)

Special Committee on Seals, *Scientific Advice on Matters Related to the Management of Seal Populations*, annual reports (1990–2019)

Stansbury, A. L. & Janik, V. M., 'Formant Modification through Vocal Production Learning in Gray Seals', *Current Biology*, 29, pp. 2244–2249 (2019)

Steel, David, *Isle of May National Nature Reserve*, blog (Feb 2015–present): https://isleofmaynnr.wordpress.com/

Bibliography

Steel, David, 'Great Bird Reserves: the Farne Islands', *British Birds*, vol. 111, issue 1, pp. 25–41 (2018)

Stetka, Bret, 'Seals Use Their Whiskers to See and Hear', *The Atlantic* (13 October 2015): https://www.theatlantic.com/science/archive/2015/10/seals-use-their-whiskers-to-see-and-hear/413269/

Studying Seals – Research at Durham University on the Behavioural Ecology of the Grey Seal in the UK, blog with multiple authors (2012–2020): https://sealbehaviour.wordpress.com/

Sylvester, Simon, *The Visitors* (London: Quercus, 2014)

Tabor, June, 'The Great Selkie of Sule Skerry' on *Ashore* [CD] (Topic Records, 2011)

Taylor, Kate, '"Shark!": Cape Cod Recoils in a Summer of Great White Sightings, Real and Imagined', *The New York Times* (20 August 2019): https://www.nytimes.com/2019/08/20/us/sharks-cape-cod-panic.html

Taylor-Beales, Rachel, *Stone's Throw – Lament of the Selkie* [CD] (Hushland, 2015)

Taylor-Beales, Rachel, *Turbulence Parts 1–10*, blog (August/September 2017): https://rachel-t-b.blogspot.com/2017/

Thompson, D., Hammond, P. S., Nicholas, K. S. & Fepak, M. A., 'Movements, diving and foraging behaviour of grey seals (*Halichoerus grypus*)', *Journal of Zoology*, vol. 224, issue 2, pp. 223–232 (1991)

Thompson, D. & Duck, C., *Berwickshire and North Northumberland Coast European Marine Site: grey seal population status*, Sea Mammal Research Unit report to Natural England: 20100902-RFQ (2010)

Thompson, D., Brownlow, A., Onoufriou, J. & Moss, S., *Collision risk and impact study: Field tests of turbine seal carcass collisions*, Sea Mammal Research Unit report to Scottish Government, no. MR 5 (2015)

Thompson, David, *Assessment of Risk to Marine Mammals from Underwater Marine Renewable Devices in Welsh Waters (on behalf of the Welsh Government), Phase 2: Studies of Marine Mammals in Welsh High Tidal Waters, Annex 1: Movements and Diving Behaviour of Juvenile Grey Seals in Areas of High Tidal Energy* (2012)

Thompson, Paul, *The Common Seal* (Aylesbury: Shire Publications, 1989)

Thomson, David, *The People of the Sea: Celtic Tales of the Seal Folk* (Edinburgh: Canongate, 2001)

Towrie, Sigurd, 'The Origin of the Selkie-folk: Documented Finmen sightings', *Orkneyjar* website: http://www.orkneyjar.com/folklore/selkiefolk/origins/origin5.htm

Trigg, L. E., Shapiro, G. I., Ingram, S. N. et al., 'Predicting the exposure of diving grey seals to shipping noise', *The Journal of the Acoustical Society of America*, 148, 1014 (2020)

Vouillemin, Eveline, 'Singing seals open new research avenues', *Geographical* (8 July 2019): https://geographical.co.uk/nature/wildlife/item/3257-singing-seals

Wallace, James, *A Description of the Isles of Orkney* (Edinburgh: William Brown, 1883; reprinted from the original 1693 edition)

Warner Hooke, Nina, *The Seal Summer* (London: Pan, 1967)

Watson, Lee, Ythan Seal Watch Facebook page (2016–present): https://www.facebook.com/ythansealwatch/

Watson, Paul, *Seal Wars* (Toronto: Key Porter Books, 2002)

Westaway, Jonathan, 'The Inuit discovery of Europe? The Orkney Finnmen, preternatural objects and the re-enchantment of early-modern science', *Atlantic Studies* (2020): https://www.tandfonline.com/doi/full/10.1080/14788810.2020.1838819

Whitworth, Victoria, *Swimming with Seals* (London: Head of Zeus, 2017)

Williamson, Duncan, *Land of the Seal People* (Edinburgh: Birlinn, 2010)

Wilson, L. J. & Hammond, P. S., 'The diet of harbour and grey seals around Britain: Examining the role of prey as a potential cause of harbour seal declines', *Aquatic Conservation,* vol. 29, issue S1, *Supplement: Sea Mammal Research Unit 40th Anniversary: Advances in Marine Mammal Science Informing Policy*, pp. 71–85 (2019)

Wilson, Susan, *Seal–Fisheries Interactions: Problems, Science and Solutions* report commissioned by British Divers Marine Life Rescue (2002)

Woolf, Virginia, *To the Lighthouse* (London: Grafton Books, 1987)

Worthington Wilmer, J., Allen, P. J., Pomeroy, P. P. et al., 'Where have all the fathers gone? An extensive microsatellite analysis of paternity in the grey seal (*Halichoerus grypus*)', *Molecular Ecology*, vol. 8, no. 9, pp. 1417–1429 (1999)

Bibliography

Worthington Wilmer, J., Overall, A. J., Pomeroy P. P. et al., 'Patterns of paternal relatedness in British grey seal colonies', *Molecular Ecology*, vol. 9, no. 3, pp. 283–292 (2000)

Wu, Katherine J., '"Talking" seals mimic sounds from human speech, and validate a Boston legend', *Nova* (26 June 2019): https://www.pbs.org/wgbh/nova/article/seals-mimic-speech/

Acknowledgements

From initial idea to published book, *Where the Seals Sing* has been hugely enriched by the knowledge, insights and thoughtful attention of so many people.

At the start of my round-Britain journey, it was a privilege to learn from stellar seal advocate Sue Sayer, both on the cliffs of West Cornwall and at Seal HQ, and Dan Jarvis at the Cornish Seal Sanctuary was the most helpful of hosts. In West Wales, the passion and commitment of Lisa Morgan and Birgitta Büche were inspiring, while Pauline Bett was an engaging seal-watching companion on her home patch of Cwmtydu and, on a reciprocal visit to Aber Felin, admirably unfazed by the ankle-deep mud. Thanks to Bel Deering for the tour of RSPCA West Hatch's seal rehab facilities and the bonus encounter with the rescued ferrets, and to both David Gregson of the Friends of Hilbre and Dr Lara Howe of Manx Wildlife Trust for the phone chats and seal survey records.

Dr Emily Doolittle offered a fascinating introduction to zoomusicology and I look forward to continuing our conversation on my future trips to Glasgow. I'm grateful to Rachel Taylor-Beales, not only for her courageous contribution to this book but also for her subsequent friendship.

Andy Ottaway of the Seal Protection Action Group, Sea Shepherd UK's Rob Read, and hunt sabs Jenny Green and Alfie Moon all

Acknowledgements

generously shared their experiences and expertise as I tackled the issue of seals–fisheries interactions. I much appreciated Lee and Eilidh Watson's warm welcome and revelations about their work as defenders of the seals of the Ythan Estuary. My thanks are also extended to David Steel for offering his unique perspective on two major breeding colonies with wisdom and wit, and to Professor Ailsa Hall who addressed an exhilarating range of issues in the short time that was available during my Sea Mammal Research Unit visit. Dr Kimberley Bennett and Dr Kelly Robinson gave an absorbing and accessible overview of the complexities of the PHATS project and, over a period of several months, Kimberley also graciously responded to all my seal physiology queries: any inaccuracies that exist in the text are wholly my own.

In England's North East, Sal Bennett furnished me with a wealth of material in the characterful settings of lighthouse and bird hide, while Ben Burville opened my eyes to the wonders of seals' underwater lives. Further south, in the midst of Donna Nook's pupping season, Rob Lidstone-Scott and Lizzie Lemon juggled reserve management issues, visitor engagement and my copious questions with aplomb.

From the Norfolk stage of my travels, gratitude is due to Friends of Horsey Seals stalwart, Gemma Walker, as well as to RSPCA East Winch's Alison Charles both for the guided tour and the exceptional levels of compassion shown to the sick and injured seals in her care. I'd particularly like to thank Ajay Tegala for allowing me to join the pup count at Blakeney and providing me with the most unforgettable wildlife experience of my seal journey, if not my life.

Many other scientists, animal care workers, rangers, artists, activists and volunteers contributed useful information while I was working on this book and, though they might not explicitly appear in its pages, they undoubtedly helped to shape it. These include John Arnott, Elizabeth Bradfield, Alex Cornelissen, Imogen Di Sapia, Simon Fathers, Sheryl Fink, Amanda James-Hussey, Dr Damian Lidgard, Kate Lock, Debbie MacKenzie, Kim Mercer, Uisdean Murray, Dr Sarah Perry, Brian Sharp, Powell Strong and the Pembrokeshire

College Wildlife Observer Wales course team, David Vyse and Dr Sue Wilson.

It's not, of course, only these human animals to whom I must pay tribute but also the more-than-human – the hundreds of grey seals in whose presence I've spent so many cherished hours and days. In the process of unobtrusively watching and listening, I've learnt so much about their behaviour and biology and I'm humbled that they consented to my sharing their shores.

For all the research, travel and fieldwork, this book wouldn't exist were it not for the keen eyes and kind heart of my agent, Charlotte Atyeo, of Greyhound Literary: she believed in it from the start and has never failed to give pitch-perfect encouragement and advice. At William Collins, my editor, Myles Archibald, supplied invaluable thematic and structural guidance, while Sally Partington – patient, creative and wise – made the copyediting process a real delight. I also feel fortunate that Hazel Eriksson, along with cover designer Ola Galewicz, Mark Bolland, Malcolm Jones, Helen Upton and Chris Wright, brought the book to completion, and sent it out into the world, with such flair and enthusiam.

Finally, three extra-special sets of thanks. To Dad, Geoff Richardson, for so many years of shared nature-love and laughter and for all the read-aloud stories in childhood and beyond. To Hooper, whose constant curiosity and tail-wagging joy amplify the pleasure of every coastal walk. And to Russell Holden, whose boundless support, both emotional and practical, nurtured me through the whole research and writing process. After a windy, wintry, cliff-top seal-watch, I couldn't wish for a finer provider of hot tea and cake.

Index

Index

Index